Particle Astrophysics and Cosmology

NATO ASI Series

Advanced Science Institutes Series

A Series presenting the results of activities sponsored by the NATO Science Committee, which aims at the dissemination of advanced scientific and technological knowledge, with a view to strengthening links between scientific communities.

The Series is published by an international board of publishers in conjunction with the NATO Scientific Affairs Division

A **Life Sciences**	Plenum Publishing Corporation
B **Physics**	London and New York
C **Mathematical**	Kluwer Academic Publishers
and Physical Sciences	Dordrecht, Boston and London
D **Behavioural and Social Sciences**	
E **Applied Sciences**	
F **Computer and Systems Sciences**	Springer-Verlag
G **Ecological Sciences**	Berlin, Heidelberg, New York, London,
H **Cell Biology**	Paris and Tokyo
I **Global Environmental Change**	

NATO-PCO-DATA BASE

The electronic index to the NATO ASI Series provides full bibliographical references (with keywords and/or abstracts) to more than 30000 contributions from international scientists published in all sections of the NATO ASI Series.
Access to the NATO-PCO-DATA BASE is possible in two ways:

– via online FILE 128 (NATO-PCO-DATA BASE) hosted by ESRIN,
Via Galileo Galilei, I-00044 Frascati, Italy.

– via CD-ROM "NATO-PCO-DATA BASE" with user-friendly retrieval software in English, French and German (© WTV GmbH and DATAWARE Technologies Inc. 1989).

The CD-ROM can be ordered through any member of the Board of Publishers or through NATO-PCO, Overijse, Belgium.

Particle Astrophysics and Cosmology

edited by

Maurice M. Shapiro

Department of Physics and Astronomy,
University of Maryland,
College Park, Maryland, U.S.A.

Rein Silberberg

Universities Space Research Association,
Washington, D.C., U.S.A.

and

John P. Wefel

Department of Physics and Astronomy,
Louisiana State University,
Baton Rouge, Louisiana, U.S.A.

Springer Science+Business Media, B.V.

Proceedings of the NATO Advanced Study Institute on
Particle Astrophysics and Cosmology
Erice, Italy
June 20–30, 1992

Library of Congress Cataloging-in-Publication Data

A C.I.P. Catalogue record for this book is available from the Library of Congress.

ISBN 978-94-010-4748-7   ISBN 978-94-011-1707-4 (eBook)
DOI 10.1007/978-94-011-1707-4

Printed on acid-free paper

TABLE OF CONTENTS

III. COSMOLOGY

PREFACE

The symbiosis between particle physics and cosmology has virtually become a conjugal relationship. Hence the 9th biennial Course of the International School of Cosmic-Ray Astrophysics was designed to bridge these formerly disparate disciplines. This NATO Advanced Study Institute (ASI) took place at the Ettore Majorana Centre in Erice, Italy, June 20-30, 1992. Seventy participants from 17 countries enjoyed the opportunities for lively interactions as much as they benefitted from the stimulating lectures. This volume is based on a selection of lectures and shorter talks presented at the sessions.

Warm thanks are due to my co-director, Prof. J. P. Wefel and to co-editor Dr. Rein Silberberg for their cooperation. The support of NATO's Scientific Affairs Division and of Dr. L. V. da Cunha, Director of its ASI Programme, was invaluable. We also acknowledge important contributions by the following: Prof. A. Zichichi, Director of the Majorana Centre and its dedicated staff; the Italian Ministry of Education; the Italian Ministry of Scientific Research; the Sicilian Regional Government; the National Science Foundation of the USA, the European Physical Society, and Mrs. Shirley Ratner of Bethesda, Maryland.

The Scientific Advisory Committee consisted of Profs. P. V. Auger, G. P. S. Occhialini, B. Rossi, M. M. Shapiro, R. Silberberg, J. A. Simpson, J. A. Van Allen, J. P. Wefel, and A. Zichichi. All of the foregoing persons and agencies helped make this ASI a memorable experience for the participants.

Maurice M. Shapiro[*]
University of Maryland

---

[*]Address for correspondence:
205 Yoakum Parkway, # 1514
Alexandria, VA 22304, USA

COMPOSITION, PROPAGATION AND REACCELERATION OF COSMIC RAYS

R. SILBERBERG
Universities Space Research Association
Washington, DC   20024

C.H. TSAO
E.O. Hulburt Center for Space Research
Naval Research Laboratory
Washington, DC 20375-5352

and J.R. LETAW
Severn Communications corporation
Millersville, MD 21108

ABSTRACT. We briefly review the composition of cosmic rays, and the modification of the source composition by nuclear spallation in interstellar space. The source composition is explained in terms of flare star particles, with rigidity dependent escape (injection) into the interstellar medium and acceleration by shock waves of relatively young Supernova remnants. Some injection from the very high velocity stellar wind particles of Wolf-Rayet stars is also required. About 500 measurements of nuclear abundances $3 < Z < 28$ have been explored from 50 to 2×10^5 MeV/n. Four models of reacceleration (with several submodels of variable reacceleration strengths) were explored; two of these are presented here. A χ^2 analysis shows that the reacceleration models yield at least equally good fits to data as the standard propagation model. However, with reacceleration, the ad hoc discontinuous assumptions of the standard model are eliminated. Furthermore, the difficulty between rigidity dependent leakage and energy independent anisotropy below energies of 10^{14} eV is alleviated.

1. Introduction

The nuclear composition and energy spectra of cosmic rays are related on the one hand to nucleosynthesis in stars, evolution of composition of stars and of the interstellar medium, and on the other hand to processes in interstellar space due to magnetic fields, hydromagnetic shock waves of young and old supernova remnants, and spallation of cosmic-ray nuclei in the interstellar gas, especially in the relatively dense clouds.

The cosmic rays play a significant role in Galactic dynamics. The energy densities of (1) cosmic rays, (2) the magnetic fields in Galaxy and (3) of the thermal gas in the interstellar medium are about 1 eV/cm^3 each. The annual energy input into Galactic cosmic rays is about 10^{60} eV/year (calculated from the cosmic-ray energy density, the volume of the Galactic radio disk and the cosmic ray confinement time in the Galaxy). The number of Galactic cosmic rays accelerated per year is about 10^{51}, with a mean energy of about 10^9 eV; (this is close to the mass of the earth per year). During the age of the Galaxy of 10^{10} years about 10^4 solar masses of Galactic particles have been accelerated to become cosmic rays.

1

M. M. Shapiro et al. (eds.), Particle Astrophysics and Cosmology, 1–16.
© 1993 *Kluwer Academic Publishers.*

2. The Composition of Cosmic Rays

The principal difference between the composition of cosmic rays and normal main sequence stars like the sun is the build-up of normally rare nuclei by spallation. About half of cosmic-ray nuclei heavier than helium have suffered nuclear collisions in the interstellar medium breaking up into lighter nuclei. For example the abundances of ^{2}H and ^{3}He are built up by the spallation of ^{4}He; Li, Be, and B are built up by the spallation of ^{12}C and ^{16}O; ^{14}N and ^{15}N are built up by the spallation of ^{16}O. ^{19}F is built up by the spallation of ^{20}Ne, ^{24}Mg and ^{28}Si; Sc, Ti, V, Cr, Mn are built up by the spallation of ^{56}Fe, and elements with $61 \lesssim Z \lesssim 75$ are built up by spallation of Pt and Pb. Figure 1 illustrates the effects of spallation with a comparison of the relative abundances of elements He to Ni in cosmic rays and in the solar system.

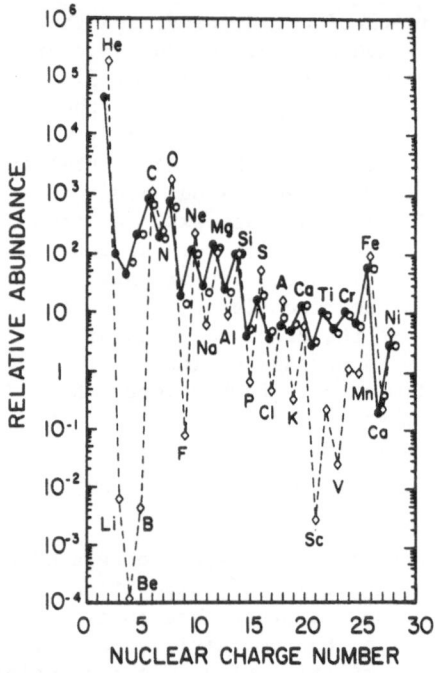

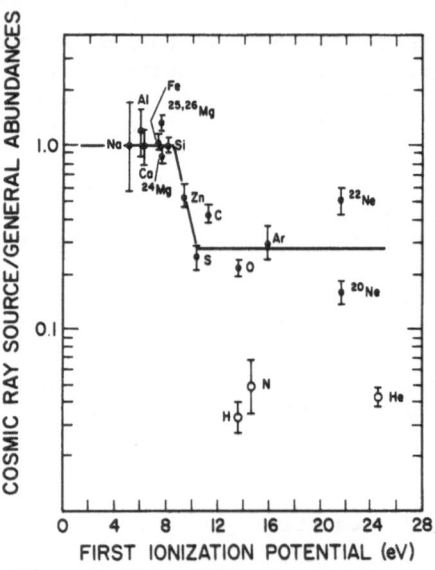

Fig. 2. The ratio of cosmic-ray source abundances to general abundances, normalized at Si (Silberberg and Tsao, 1990).

Fig. 1. The cosmic ray element abundances (He-Ni) measured at earth compared to solar system abundances, all relative to silicon. Solid circles: low energy data, 70-280 MeV/n, open circles: compilation of high energy data, 1000-2000 MeV/n; diamonds: solar system. The figure is based on Garcia-Munoz and Simpson (1979) and Simpson (1983).

With the help of spallation cross sections and the observed cosmic-ray composition, one can calculate the source composition of cosmic rays. In this calculation, one must take into account the uncertainties in cosmic-ray composition and of spallation cross sections. The abundances of elements Li to Ni are now known to a precision of $\lesssim 5\%$, due to measurements by many groups, especially Engelmann et al. (1990). The cross sections for nuclear stripping and spallation could be calculated to $\sim 30\%$ using the semiempirical calculations of Silberberg and Tsao

(1973), whose subsequent refinements have reduced the error to ~ 20%. Recently, Webber et al. (1990) have measured cross sections with a high degree of precision, and constructed semiempirical equations that for a given set of product isotopes yield a fit to about 10%.

As was shown in the previous Erice proceedings by Silberberg et al. (1991), that the source abundances of cosmic rays are strikingly similar to the general abundances of elements and isotopes. The general abundances of normal main sequence stars are taken to be that of the sun with isotopic abundances for the non-volatile elements adopted from the carbonaceous chondrite meteorites). While the general abundances vary by a factor of 10^{10} as one proceeds from H to Pb, the ratio of cosmic ray source abundances to the general abundances agrees within a factor of 5 except for H, He, and N, which are underabundant in cosmic rays by a factor of about 30. This is illustrated in Fig. 2.

Fig. 2 shows that elements whose first ionization potential (FIP) exceeds 10 eV are about 4 times less abundant in cosmic rays. Furthermore, the light elements H, He, and N are underabundant by an additional factor of 8. Finally, we note a dichotomy of the ratios $^{22}Ne/^{20}Ne$, $^{25,26}Mg/^{24}Mg$ and C/O. Elements with FIP > 10eV are the noble gases He, Ne, Ar, Kr, and Xe, the gases H, N, O, the halogens F, Cl, Br, and I, and C, P, S, and Hg. The physical meaning of FIP is: if the energy input to an atom equals or exceeds slightly the FIP, the atom can be stripped of an electron from its outermost filled energy level producing a singly ionized atom. The temperature at which such ionization occurs is like that of the photosphere of a normal star like the sun. The solar coronal composition and that of the solar wind and flare particles derived from the corona displays a dearth by a factor of 4 of elements with FIP > 10 eV, just like the cosmic rays. The FIP-dependent suppression of certain elements in cosmic rays was first proposed by Havnes (1971) and a detailed investigation was carried out by Casse, Goret, and Cesarsky (1975). The difference between the photospheric and coronal composition is suggestive of diffusion from the photosphere into the corona, with easier diffusion for the singly ionized nuclei than for the neutral nuclei, possibly along magnetic field lines. This suggests that the cosmic rays are derived from the outermost layers of stars. Meyer (1985) points out that a further stage of injection (stellar flare particles near the shock waves of supernova remnants) at energies near 1 MeV/nucleon is plausible as relative ionization loss effects on particle range and composition cancel due to the effective charges of atoms near energies of 1 MeV/n. Shapiro (1990) presents several strong arguments that the very numerous flare stars are likely to be the injectors of cosmic rays.

The dearth of H, He, and N can be explained by the model of Silberberg and Tsao (1990). It is presented in the previous Erice proceedings by Silberberg et al. (1991), which is repeated here for the sake of completeness of our explanation of the differences between the source composition of cosmic rays and the general elemental and isotopic abundances.

The explanation proposed here would result in an underabundance of C and O. However, this is compensated by nucleosynthesis in WC stars, which is discussed in the next subsection. The underabundance could be due to a rigidity cutoff or deceleration at the boundaries of astrosphere (analogous to the heliosphere) and interstellar space, due to the effective charge dependence of the rigidity. (Rigidity is proportional to the radius of curvature of a particle in a magnetic field). Near 1 MeV, multiple electron pickup, with $(Z_{eff}/Z) < 1$ becomes appreciable in the interval C to Mg. (The effective charge of a particle is the nuclear charge minus the number of electrons attached to the nucleus). In the discussion below we make use of the following relationship: rigidity = momentum/Z_{eff} = (momentum/nucleon) x (A/Z_{eff}). From Northcliffe and Schilling (1970; Table and eq. on page 236), one can deduce that the ratio of mass number to effective charge (near energies of 1 MeV per nucleon) A/Z_{eff} = 2.0 for He and 2.95 for Mg. Thus, if there is an effective rigidity cutoff R_0 near the boundary of the astrosphere, in the neighborhood of energies of 1 MeV/n, the momentum per nucleon at the cutoff P_0 = R_0 x Z_{eff}/A = 0.5 R_0 for He and 0.339 R_0 for Mg, i.e., the momentum per nucleon at the cutoff is numerically less than the cutoff rigidity, expressed in same units, MV/c. Thus, a greater part of the momentum spectrum of Mg passes the cutoff than of the spectrum of He. If the exponent of the integral momentum spectrum of stellar flare particles is 5, i.e., $J(P>P_0) \alpha P_0^{-5}$, then He is suppressed relative to Mg by a factor of $(2.95/2)^5 = 7$. The exponent of the stellar flare spectra is chosen to be 5, to yield a suppression of He and N relative to Mg and Si by a factor of 7 or 8. The suppression of H relative to He probably has a velocity dependent component, as discussed by Silberberg and Tsao (1990).

The near-constancy of the ratio of the cosmic-ray source abundance to the general abundance above $Z \gtrsim 12$ would require a "bending over" or a threshold in the momentum per nucleon spectrum of flare particles, so that nuclei with momentum per nucleon values smaller than those of $Z = 12$ at the astrospheric rigidity threshold are nearly excluded. (At the value near 1 MeV for flare particles the relative ionization loss effects on the range of particles nearly cancel due to pickup of electrons by the nuclei, as shown by Meyer, 1985).

We note from Fig. 2, that in addition to H, He, and N, the nuclide ^{20}Ne can be interpreted as being affected by the process of suppression of light nuclei. This implies that C and O in the principal cosmic-ray source are affected. We also note a significant discrepancy in the ratio of the cosmic-ray abundances to the general abundances of C and O, see Fig. 2, (removed by the combined effects of light-ion suppression and Wolf-Rayet star contributions). The determination of the suppression factor from first principles would require detailed knowledge of the energy spectrum of stellar flares and of processes at the astrosphere boundary shocks. In the absence of such knowledge, we shall adopt an empirical factor S(Z), where $0.12 \leq [S(Z) = 0.15 Z - 0.93] \leq 1.0$, i.e., S(Z) = 0.12 for $Z \leq 7$ and 1 for $Z \gtrsim 13$. This factor explains the values of H, He, N, and ^{20}Ne in Figure 2. The ratio of cosmic-ray to general abundances is given by I S(Z), except for the nuclides with a major component from Wolf-Rayet stars. The term I = 0.27 is the factor that represents the correction for FIP, if FIP$\gtrsim$10.4 eV. I = 1 for FIP $\leq$ 8.4, and there is a transition region in between, shown by the line in Figure 2, and the element Zn.

A third difference between the cosmic-ray source composition and the general abundances is the enhancement of ^{22}Ne, ^{25}Mg, and ^{26}Mg. Meyer (1981) explained this in terms of a 2% contribution of Wolf-Rayet stars to the cosmic-ray source material. Prantzos et al. (1985) showed that the abundance ratio C/O in cosmic rays (about twice the solar one) can be explained in terms of Wolf-Rayet star contribution to carbon. (The winds of Wolf-Rayet stars are energetic, close to 0.1 MeV per nucleon, relatively close to the energies of flare particles. The acceleration of these wind particles thus is plausible). The Wolf-Rayet stars go through two phases: WN, when N produced in the CNO hydrogen-burning cycle is abundant at the stellar surface, and WC, when C, produced in helium burning is abundant at the surface, and ^{14}N burns into ^{22}Ne.

However, we suggest a modification of the calculations of Prantzos et al. (1985), that reduces the ratio of $^{22}Ne/(C + O)$, or enhances C and O, compensating for light-ion suppression. The calculated abundance (mass fraction) for ^{22}Ne of Prantzos et al. (1986) is 1.6 times higher than that of Maeder (1983, 1987) and Maeder and Meynet (1987). This is because Prantzos et al. (1986) assume the mass fraction $(Z > 2)/(Z \leq 2) = 0.03$, while the general solar system abundances of Grevesse and Anders (1989) yield 0.019 and of Cameron (1982) yield 0.018, in agreement with the latter. (The value of Prantzos et al. (1986) probably is good for the Galactic center region that has a higher metallicity and a large concentration of Wolf-Rayet stars). Hence, the initial CNO abundances of Prantzos et al. (1986) should be reduced by $0.019/0.03 = 0.63$. Thus, the abundance of N for the WN phase calculations of Prantzos et al. should be reduced by 0.63, and also of ^{22}Ne and ^{25}Mg and ^{26}Mg during the WC phase. The abundances of C and O during the WC phase are not thus affected, since these are formed from He.

To sum up, this model assumes a rigidity dependent escape of flare-star particles from the astrospheres (analogous to the heliosphere) that are not fully ionized for $Z \gtrsim 6$. Some contribution from the Wolf-Rayet stars especially to C, O, ^{22}Ne, ^{25}Mg, ^{26}Mg is assumed. With this model, the calculated source abundances agree to 20% with those deduced from the observed "arriving" abundances, for elements $1 \leq Z \leq 28$ and their isotopes, if measurable above the secondary background.

3. Cosmic-Ray Propagation Equation

Figure 1 of a previous section demonstrated the importance of nuclear transformation in understanding the composition and the arriving composition. The procedures of cosmic-ray propagation calculations have been described by Ginzburg and Syrovatskii (1964), and the particular methods used in our group by Letaw et al. (1983). In collaboration with Eichler, the procedures of Eichler (1980) for a modest degree of reacceleration were introduced into our propagation calculations.

The updated transport equation of cosmic rays (also referred to as the diffusion equation or propagation equation) is given below. The equation assumes that nuclear fragments maintain the velocity of their progenitors, and that the interstellar medium is approximated by hydrogen. (The effects of 10% helium can be treated in a first approximation by a scaling factor.)

$$\frac{\partial J_i(E)}{\partial X} = \frac{Q_i(E)}{\rho} - r_e(E)_i J_i(E) + \frac{\partial}{\partial E}[W_i(E)J_i(E)]$$

$$- r_a(E)J_i(E) + (\gamma-1)v^{-1}p^{-\gamma}\int_0^E r_a(E)p^{\gamma-1}J_i(E)dE$$

$$- r_f(E)_i J_i(E) + \sum_j r_f(E)_{ij}J_j(E)$$

$$- r_d(E)_i g_i(E)J_i(E) + \sum_j r_d(E)_{ij}g_j(E)J_j(E)$$

Here Q is the injection rate (particles/cm^3-sec-MeV), ρ is the density, r_e is the escape rate (cm^2/g), also proportional to $(A_i p / Z_i)^{-\alpha}$, where α is the index of rigidity dependent escape and p is the momentum per nucleon. W is the stopping power (MeV/g-cm^{-2}-nucleon). The inverse mean free path-length for acceleration is r_a(cm^2 g^{-1}). The shock reacceleration index is γ; it equals the exponent of the momentum spectrum to which a δ-function momentum spectrum is accelerated. Fragmentation terms contain $r_f(E)_i = N_A \sigma_i(E)/A_H$, and $r_f(E)_{ij} = N_A \sigma_{ij}(E)/A_H$, the inverse mean free paths for spallation. Here N_A is the Avogadro number, σ_i the total inelastic cross section of nuclide i and σ_{ij} is the cross section for nuclide j going into i. A_H is the isotopic weight of hydrogen, $r_d(E)_i = (1+E/m)^{-1} N_A(\ln 2)/(n_H A_H v \tau_i)$ and $r_d(E)_{ij} = (1+E/m)^{-1} N_A(\ln 2)/(n_H A_H v \tau_{ij})$ are the inverse mean free paths for decay, where m is the atomic mass unit (931.5 MeV), n_H = number density of hydrogen in the transport medium, v = velocity (cm/sec), τ_i = half life of species i and τ_{ij} is the half life of species j decaying into i, in units of seconds, $g_i(E) = 1$, except for the nuclei that decay by electron capture; then $g_i(E) \simeq r_+(E)_i/(r_{ec}(E_i)+(r_-(E_i)))$, where $r_{ec}(E_i)$, the inverse mean free path for decay by electron capture is of the same form as $r_d(E)_i$ above, the r_+ and r_- are the inverse mean free paths for electron attachment and stripping, of same form as $r_f(E)_i$ above.

The complete network of calculations includes all the isotopes, (several hundred), and cross sections of all target-product nucleon pairs, several thousand in number.

4. Modification of Energy Dependence of Cosmic-Ray Abundance Ratios by Radioactive Decay.

There are several long-lived ($\tau_{1/2} > 10^5$ years) radioactive isotopes in cosmic rays: ^{10}Be, ^{26}Al, ^{36}Cl and ^{54}Mn and several more trans-iron nuclides. In addition, there are isotopes that decay by electron capture, after the attachment of an electron to a bare nucleus. For elements $8 < Z < 25$, the nuclei below E ~ 300 MeV/n are thus affected.

Below, some figures demonstrate the effects of decay, and the suppression of decay by the time-dilation Lorentz factor above energies of GeV/n. Figure 3 shows the ratio Be/B, modified by the decay of ^{10}Be. The dashed line shows the ratio for no decay, and the solid curve for decay in a medium with density 0.3 atoms/cm^3. The dotted line shows the effect of solar modulation of ϕ = 300 MV. The circles are the data of Engelmann et al. (1990), the data point near 200 MeV/n is from Garcia-Munoz and Simpson (1979), and the low-energy point is from Ferrando et al. (1991). Figure 4 shows the corresponding curves for no decay, decay and solar-modulation of ^{10}Be/^{9}Be, where the triangle is the data point from Wiedenbeck and Greiner (1983) and the rectangle from Garcia-Munoz et al. (1979). Fig. 5 shows the ratio ^{38}Ar/^{36}Ar with the same curves. Below 2 GeV/n, the decay of ^{36}Cl nearly doubles the abundance of ^{36}Ar, thus reduces ^{38}Ar/^{36}Ar by a factor of 2. The data point of Webber (1981) is consistent with the decay of most ^{36}Cl below a few GeV/n. Fig. 6 shows the effects of decay of ^{49}V that decays by electron capture at energies of several 100 MeV/n. The data point of Webber (1981) has too large an error to draw any conclusion except for the need of future experiments with both good statistics and good mass resolution.

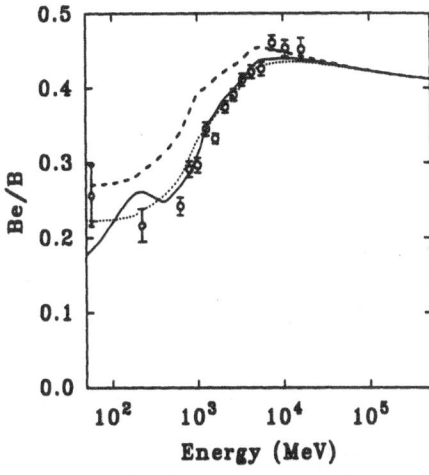

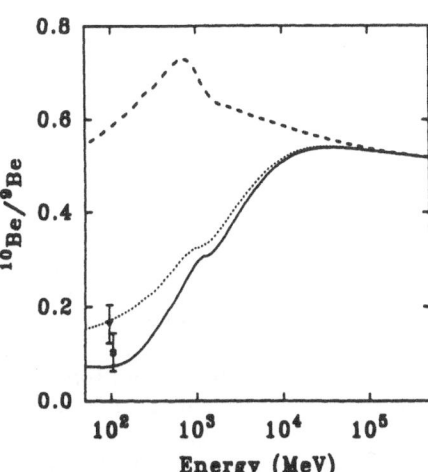

Fig. 3. Ratio of Be/B as a function of energy. The solid curve is with no solar modulation, the dotted one with a modulation of 300 MV, and the dashed curve shows the ratio if ^{10}Be did not decay.

Fig. 4. Ratio of ^{10}Be/^{9}Be as a function of energy. The curves have been defined Fig. 3.

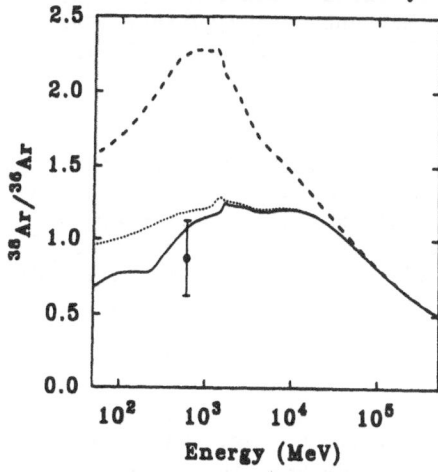

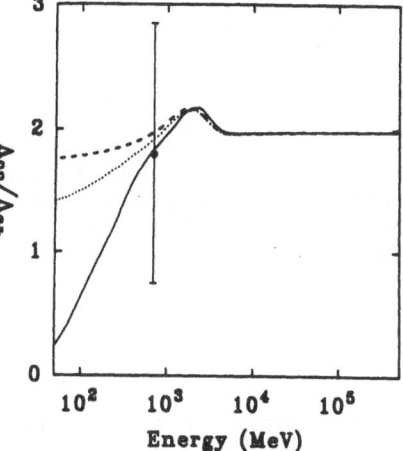

Fig. 5. Ratio of ^{38}Ar/^{36}Ar as a function of energy. The dashed curve is for the case of no ^{36}Cl decay.

Fig. 6. Ratio of ^{49}V/^{50}V as a function of energy. The dashed curve is for no decay of ^{49}V.

5. Distributed Reacceleration

The principal acceleration of cosmic rays probably is carried out by shock waves of relatively young supernova remnants. Some reacceleration by old, weakened shock waves that permeate much of interstellar space must occur as shown by Blandford and Ostriker (1980), Silberberg et al. (1983) Simon et al. (1986), Cesarsky (1987), Giler et al. (1987) and Ptuskin (1991). Constraints on the degree of reacceleration have been discussed by Cowsik (1986).

We have explored the model of distributed acceleration (e.g. Wandel et al. 1987) with the intention of reducing or eliminating the fitting parameters (1) and (2) of the standard model of cosmic-ray propagation. (1) The mean path length in the standard model has a rigidity dependence α $R^{-\epsilon}$ (where ϵ= 0.5 or 0.6), related to easier escape of high energy nuclei from the Galaxy. However, at rigidity R < 5 GV, there is a discontinuity, with a reduction of the mean path length, represented by the velocity β. The reason for this discontinuity is the corresponding discontinuity in the secondary to primary abundance ratio, e.g. B/C, displayed subsequently. (2) The shock wave acceleration process results in a power law in momentum (or rigidity). After correction of energy spectra for solar modulation, one needs a reduced intensity below 5 GV. Several forms for the reduction have been proposed in our fitting procedure for the standard model; the introduction of multiplicative factor β^{κ}, with κ = 1.18 yields a good fit.

In addition, a mean path length α $R^{-\epsilon}$ with ϵ > 0.5 implies a rapidly increasing leakage rate (with rigidity) of cosmic rays from the Galaxy, which is difficult to reconcile with the low and nearly constant anisotropy of cosmic rays at energies below 10^{14} eV. Distributed acceleration permits the use of a smaller value of ϵ.

We have collected about 500 elemental and isotopic abundances from recent publications: Engelmann et al. (1990), Garcia-Munoz and Simpson (1979), Garcia-Munoz, Simpson, and Wefel (1981), Binns et al. (1988), Webber (1981), Gupta and Webber (1989), Wiedenbeck and Greiner (1980), Krombel and Wiedenbeck (1988), Wiedenbeck (1985), Webber et al. (1985), Ferrando et al. (1987), Ferrando et al. (1991), Swordy et al. (1990) and Muller et al. (1991).

Letaw, Silberberg and Tsao have optimized the propagation parameters for the standard model: (1) reacceleration with constant (i.e. continuous) power laws in momentum and the mean path length and with energy independent reacceleration; (2) same as (1) but energy-dependent reacceleration; (3) same as (2) but introducing a factor β^{κ} into the source spectrum and (3A) like (3) but using a distribution of reacceleration spectra, based on Axford (1981). Models (2), (3), (3A) yield equally good or slightly better χ^2 values than the standard model. The χ^2 values (Table 2) exceed 500, because the errors in abundance ratios due to uncertainties in cross sections (about 10%) exceed the errors of about 3% reported by Engelmann et al. (1990), which as a result of internal continuity checks was raised to ~5%. Thus the values of χ^2 quoted are not physically meaningful, however, the relative values give a measure for the relative goodness of fit of the various models. A detailed presentation for the reacceleration models (1) to (3), each with 7 sub-sets will be presented in a paper by Letaw, Silberberg, and Tsao (1992 or 1993). Here we present a comparison of model (3A) with the standard model (i.e. with a distribution of reaccelerated values based on Axford (1981), and model (2) with a reaccelerated spectrum with an exponent γ=7 for the momentum spectrum (p^{-7}). The best-fit reacceleration thus is weaker than that we presented earlier, Letaw et al. (1987), where p^{-4} was used. The distribution of reaccelerated values of model (3A) is presented below:

Table 1. Weighting factors of exponent γ

γ	2.3	3.0	4.0	5.0	6.0	7.0	8.0
Weighting factor (%)	2	8	10	10	10	10	50

The values for $\gamma > 8$ have been included into $\gamma=8$, since the effects of reacceleration beyond $\gamma=8$ are exceedingly weak. The small value of strong reacceleration of 2% is consistent with the range 0.3% to 10% estimated by Wandel (1990).

Table 2 below presents the propagation parameters of the 3 models discussed in this paper.

Table 2. Propagation parameters

Model	Γ	κ	λ_e	ϵ	R_c	ν	λ_a	α	χ^2
Standard	-2.36	1.18	5.64	-.562	5.3	0.97	--	--	702
2 ($\gamma=7$)	-2.36	--	4.26	-.452	--	--	5.7	0.39	686
3A	-2.36	1.10	4.06	-.492	--	--	12.0	0.7	686

Here the spectrum at the source is $p^\Gamma \beta^\kappa$, $\lambda_e (g/cm^2)$ is the escape mean free path from the Galaxy at rigidity 10 GV, the escape mean free path is $\lambda_e (R/10)^\epsilon \beta^\nu$; for the standard model below $R_c = 5.3$ GV it is $\lambda_e (R_c/10)^\epsilon \beta^\nu$ and $\lambda_a (R/10)^\alpha$ is the acceleration mean free path. The smallest $\chi^2=669$ was obtained with the distributed acceleration model (3), with $\gamma=7$.

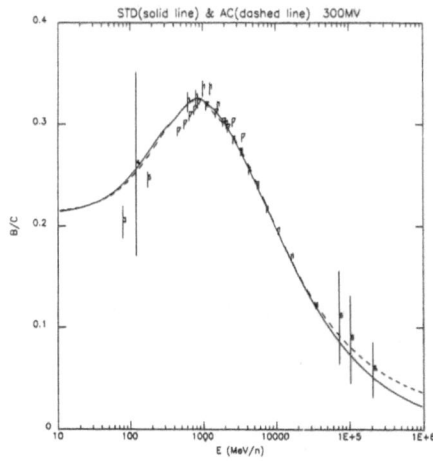
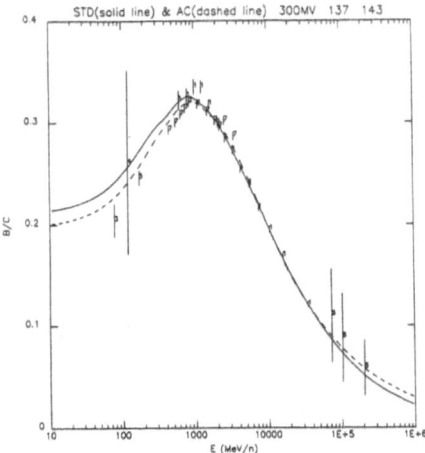

Fig. 7. Ratio B/C as a function of energy. The solid curve represents the standard propagation model and the dashed curve the reacceleration model 3A, with the propagation parameters defined in Tables 1 and 2. A solar modulation of 300 MV is assumed.

Fig. 8. Ratio B/C as a function of energy. The solid curve represents the standard propagation model and the dashed curve the reacceleration model 2 of Table 2.

In the subsequent figures various abundance ratios are displayed from 10 to 10^6 MeV/n. The solar modulation used in the calculation is ϕ = 300 MV. The experimental data, the standard model (the solid curve) and the distributed reacceleration model (the dashed curve) are intercompared. Unless otherwise stated, the reacceleration model (3A) is used. Fig. 7 shows the ratio B/C. We note the general agreement between the standard and reacceleration models. However, at 10^6 MeV/n the predicted ratio of the reacceleration model is about 50% higher. We also note that the experimental data at 600 to 3000 MeV/n of Engelmann et al. (1990) and Gupta and Webber (1989) differ by about 3 standard deviations. Such systematic differences raise the value of the χ^2. Fig. 8 shows similar data for B/C, however, instead of a distribution of exponents γ of the reaccelerated spectrum (including 2% strong reacceleration), this figure is for model (2), with the exponent $\gamma = 7$. In this model, the predicted excess with reacceleration at 10^6 MeV/n is less, 30%, instead of 50%, however, a slightly smaller B/C ratio below 1000 MeV/n, (about 5 to 10% smaller) is predicted with reacceleration.

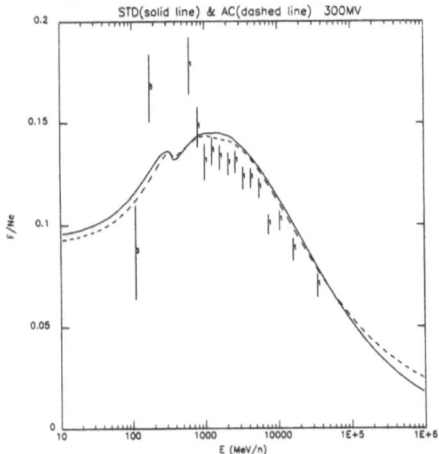

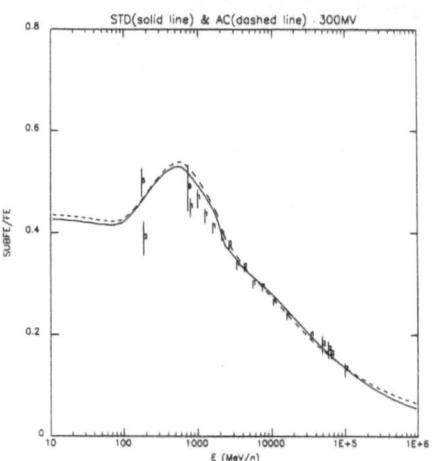

Fig. 9. Ratio F/Ne as a function of energy. The curves are defined in Fig. 7.

Fig. 10. Ratio (Sc-Cr)/Fe as a function of energy. The curves are defined in Fig. 7.

Fig. 9 compares the standard model and model (3A) for F/Ne, also a secondary-to-primary ratio. Similar small systematic differences as for B/C are seen here. The experimental values suggest that the calculated cross section for the production of F is too large by about 10%. Fig. 10 shows the ratio (Sc to Cr)/Fe for the standard model and model (3A). The excess near 10^6 MeV/n is less here (about 20%) with reacceleration, because Fe has a larger cross section than C and O, thus the competitive effects between spallation and reacceleration disfavor reacceleration.

Fig. 11 shows also the ratio (Sc to Cr)/Fe, but here model (2) with $\gamma=7$ is used, as was done in Fig. 8. Here the secondary-to-primary ratio below 10^3 MeV/u is somewhat less (5-10%) with reacceleration. Fig. 12 shows the ratio of two secondary elements K/Cr. This is interesting, because the mass difference Fe-K is large, of Fe-Cr small. Thus, below 1 GeV/n the production cross section of K decreases with decreasing

energy while that of Cr increases. Hence, also the tertiary (third
generation spallation) contribution to K becomes important at low
energies. Accordingly, with reacceleration there is a shift in the K/Cr
curve (below 1 GeV/n) toward higher energies. The reasons are: the low
value of K due to low cross section becomes shifted by reacceleration,
the large value of Cr due to large low-energy cross section is shifted,
and K, with its large tertiary contribution has a longer path length
(with its progenitors path length), hence suffers more reacceleration.

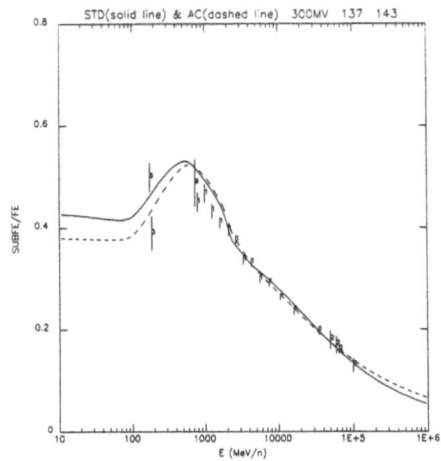

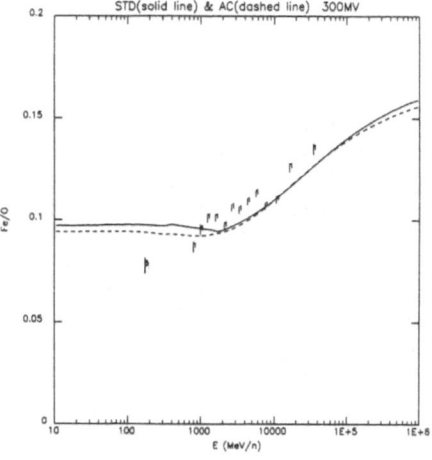

Fig. 11. Ratio (Sc-Cr)/Fe as a
function of energy. The curves
are defined in Fig. 8, i.e.
reacceleration model 2 of Table 2.

Fig. 12. The ratio K/Cr as a
function of energy, both mainly
breakup products of Fe. The curves
are defined in Fig. 7.

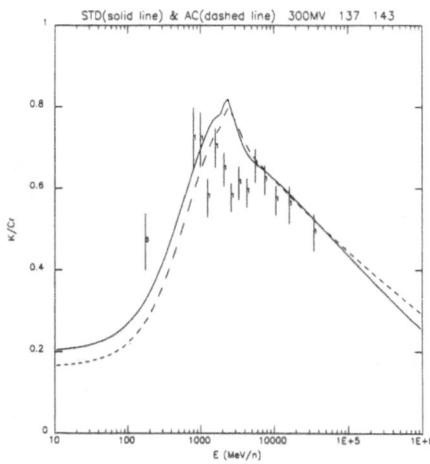

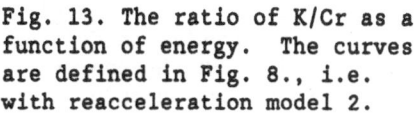

Fig. 13. The ratio of K/Cr as a
function of energy. The curves
are defined in Fig. 8., i.e.
with reacceleration model 2.

Fig. 14. The ratio Fe/O (both are
primary elements) as a function
of energy. The surves are defined
in Fig. 7.

Fig. 13 also shows the ratio K/Cr for model 2 (with $\gamma=7$). Here the shift is somewhat larger, because the acceleration frequency ($\alpha \ 1/\lambda_a$) is larger. Fig. 14 shows the ratio Fe/O, two primary nuclides. We note that the high precision of the data of Engelmann et al. (1990) can be used to improve source abundances: a 5% increase in the calculated Fe/O ratio is needed for a good fit to the data. Above 1 GeV/n the ratio increases with energy because at lower energies relatively more iron breaks up due to its large spallation cross section. Below 1 GeV/n the value Fe/O with reacceleration is somewhat lower, because before reacceleration, Fe was subject to larger ionization losses, which are larger at low energies.

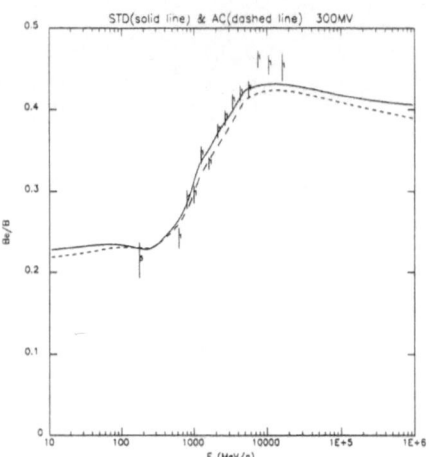 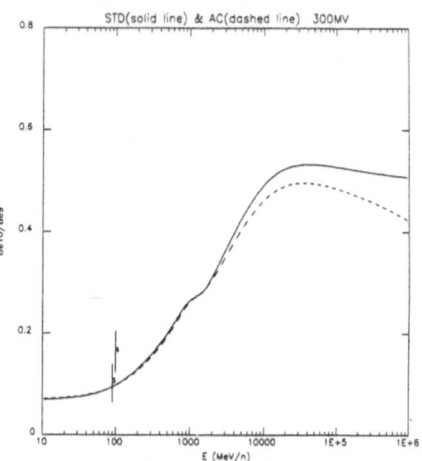

Fig. 15. The ratio Be/B as a function of energy; partial decay of ^{10}Be in a medium of density 0.3 atoms/cm^3 has been included. The curves are defined in Fig. 7.

Fig. 16. The ratio ^{10}Be/^{9}Be. Decay of ^{10}Be as in Fig. 15. The curves are defined in Fig. 7

Fig. 15 shows the ratio Be/B, calculated with partial ^{10}Be decay in a medium with a density of 0.3 atoms/cm^3. The calculation with reacceleration is slightly lower because after decay there is less ^{10}Be to be accelerated to higher energies. In this graph, the standard model appears to give a better fit to the data, but when one folds in a 10% error in cross section estimates, this conclusion need not be valid. Fig. 16 shows the ratio ^{10}Be/^{9}Be, and Fig. 17 shows the same ratio for model 2, with $\gamma=7$. The lower value in Fig. 16 for the reaccelerated curve above 10^4 MeV/n probably is due to the contribution of values $\gamma=2.3$ and $\gamma=3$ (i.e. strong reacceleration) that shifts up in energy the part of the energy spectrum that is depleted in ^{10}Be. Fig. 18 shows the ratio ^{36}Cl/^{35}Cl, where ^{36}Cl is a radioactive isotope. We note the resemblance of the curves to those of Fig. 16. The same explanation of the depletion of radioactive nuclide above 10^4 MeV/n applies.

Fig. 19 shows the ratio of ^{49}V/^{50}V, both secondary products of Fe. However, ^{49}V decays by electron attachment and capture at low energies, starting at energies near 1 GeV/n. The calculated curve is shifted up with reacceleration by about 10% because the sections in the energy spectrum with ^{49}V depletion become accelerated to higher energies.

Fig. 20 also shows the ratio ^{49}V/^{50}V, but for model 2 with γ=7. Both Fig. 19 and 20 are similar.

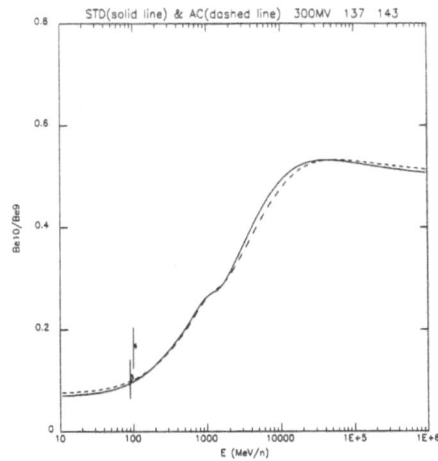

Fig. 17. The ratio ^{10}Be/^{9}Be as a function of energy. The curves are defined in Fig. 8, i.e. with reacceleration model 2.

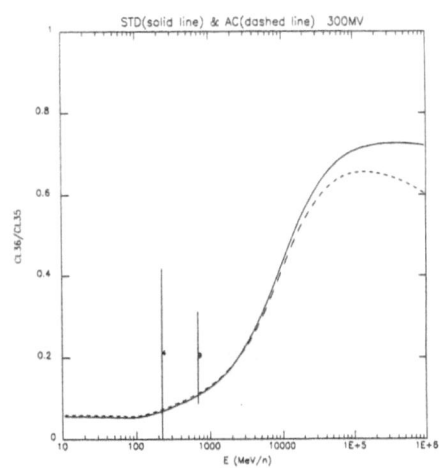

Fig. 18. The ratio ^{36}Cl/^{35}Cl as a function of energy. Partial decay of ^{36}Cl in a medium of 0.3 atoms/ cm^3 has been included. The curves are defined Fig. 7.

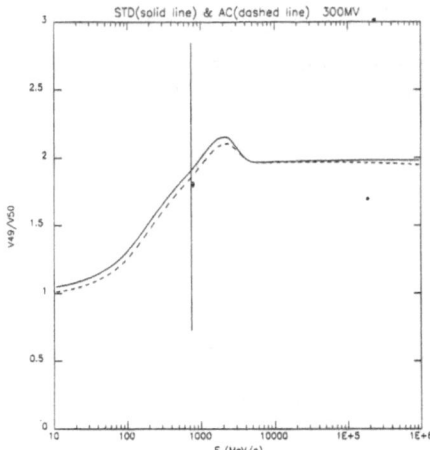

Fig. 19. The ratio ^{49}V/^{50}V as a function of energy. The curves are defined in Fig. 7. ^{49}V decays at low energies by electron capture.

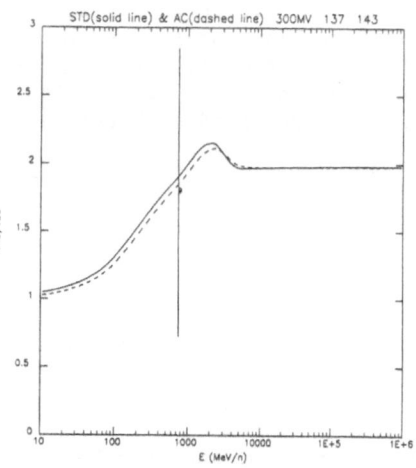

Fig. 20. The ratio ^{49}V/^{50}V as a function of energy. The curves are defined in Fig. 8, i.e. reacceleration model 2.

Fig. 21 shows the ratio ^{51}Cr/^{52}Cr, where ^{51}Cr decays by electron capture. It is similar to Fig. 19.

14

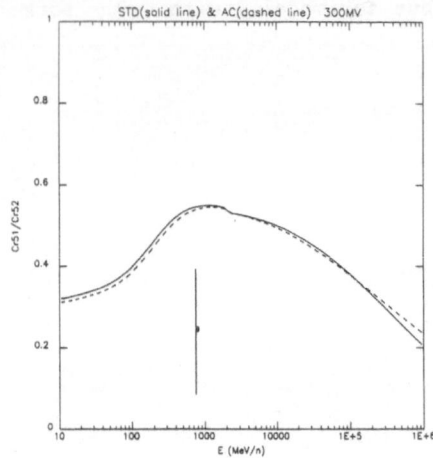

Fig. 21. The ratio $^{51}Cr/^{52}Cr$ as function of energy. The curves are defined in Fig. 7. ^{51}Cr decays by electron capture at low energies.

6. Conclusions on Reacceleration

Weak shock waves from old and/or broken-up supernova remnants permeate interstellar space and cause some reacceleration. The χ^2 calculations with reacceleration are equally good or slightly better than with the standard model. The principal justification for reacceleration is the elimination of ad hoc discontinuities in the standard model in the energy dependence of the mean path length and in the momentum spectrum at the source. Furthermore, the difficulty of energy independence of anisotropy and energy dependence of escape from the Galaxy is alleviated. Finally a test is suggested: A reduced energy dependence in secondary to primary ratios like B/C above energies of 10^6 MeV/n. This test was suggested by Wandel (1990).

ACKNOWLEDGEMENTS

This work is supported by ONR and NASA (W17,681).

REFERENCES

Axford, W.I. 1981 17th Internat. Cosmic Ray Conf. (Paris) 12, 155.
Binns, W.R., Garrard, T.L., Israel, M.H., Jones, M.D., Kamionkowki,
    M.P., Klarmann, J., Stone, E.D. and Waddington, C.J. 1988 Ap.J. 324,
    1106.
Blandford, R.D. and Ostriker, J.P. 1980 Ap.J. 237, 793.
Casse, M., Goret, P. and Cesarsky, C.J. 1975 14th Internat. Cosmic Ray
    Conf. (Munich) 2, 646.
Cesarsky, C.J. 1987 20th Internat. Cosmic Ray Conf. (Moscow) Rapporteur
    Papers 7, 87.
Cowsik, R. 1986 Astron. Astrophys. 155, 344.
Eichler, D. 1980 Ap.J. 237, 809.
Engelmann, J.J., Ferrando, P., Soutoul, A., Goret, P., Juliusson, E.,
    Koch-Miramond, L., Lund, M., Masse, P., Peters, B., Petrou N. and
    Rasmussen, I.L. 1990 Astron. Ap. 233, 96.
Ferrando, P., Engelmann, J.J., Goret, P., Koch-Miramond, L., Petrou, N.,

Soutoul, A., Herrstrom, N.Y., Byrnak, B., Lund, M., Peters, B., Rasmussan, I.L., Rotenberg, M., Westergaard, M. J. 1988 Astron. Astrophys. 193, 69.

Ferrando, P., Lal, N., McDonald, F.B., Webber, W.R. 1991 Astron. Astrophys. 247, 163.

Garcia-Munoz, M. and Simpson, J.A. 1979 16th Internat. Cosmic Ray Conf. (Kyoto) 1, 270.

Garcia-Munoz, M. and Simpson, J.A. 1979 16th Internat. Cosmic Ray Conf. (Kyoto) 1, 310.

Giler, M., Osborne, D.L., Szabelska, B., Wadowczyk J. and Wolfendale, A.W. 1987 20th Internat. Cosmic Ray Conf. (Moscow), 2, 214.

Ginzburg, V.L. and Syrovatskii, S.I. 1964. The Origin of Cosmic Rays, publ. MacMillan, N.Y.

Grevesse, N. and Anders, E. 1989 in "Cosmic Abundances of Matter", ed. C.J. Waddington (New York: AIP Conf. Proc. 183), p. 1.

Gupta, M. and Webber, W.R. 1989 Ap.J. 334, 1124.

Guzik, T.O., Wefel, J.P., Garcia-Munoz, M. and Simpson, J.A. 1985 19th Internat. Cosmic Ray Conf. (La Jolla), 2, 76.

Havnes, O. 1971 Nature 229, 548.

Krombel, K.E. and Wiedenbeck, M.E. 1985 19th Internat. Cosmic Ray Conf. (La Jolla) 2, 92.

Krombel, K.E. and Wiedenbeck, M.E. 1988 Ap.J. 328, 940.

Letaw, J.R., Silberberg, R. and Tsao, C.H. 1984 Ap.J. Suppl. 56, 369.

Letaw, J.R., Silberberg, R., Tsao, C.H., Eichler, D., Shapiro, M.M. and Wandel, A. 1987 20th Internat. Cosmic Ray Conf. (Moscow) 2, 222.

Letaw, J.R., Silberberg, R. and Tsao, C.H. to be publ. 1992 or 1993.

Meyer, J.P. 1985 Ap.J. Suppl. 57, 173.

Muller, D., Swordy, P.S., L'Heureux, J. and Grunsfeld, J.M. 1991 Ap.J. 374, 356.

Northcliffe, L.C. and Schilling, R.F. 1970 Nuclear Data Tables A7, 233.

Prantzos, N., Doom, C., Arnould, M. and de Loore, C. 1986 Ap.J. 304, 695.

Prantzos, N., Arnould, M., Arcoragi, J.P. and Casse, M. 1985 19th Internat. Cosmic Ray Conf. (La Jolla) 3, 167.

Ptuskin, V.S. 1991 in "Cosmic Rays, Supernovae and the Interstellar Medium", p. 119, ed. M.M. Shapiro et al., Kluwer Acad. Publ., Holland.

Shapiro, M.M. 1990 21st Internat. Cosmic Ray Conf. (Adelaide), 4, 8.

Silberberg, R. and Tsao, C.H. 1973 Ap. J. Suppl. 25, 315.

Silberberg, R., Tsao, C.H., Letaw, J.R. and Shapiro, M.M. 1983 Phys. Rev. Lett. 51, 1217.

Silberberg, R. and Tsao, C.H. 1990 Ap.J. Letters, 352, L49.

Silberberg, R., Tsao, C.H., Shapiro, M.M. and Biermann, P.L. 1991 in "Cosmic Rays, Supernovae and the Interstellar Medium", p. 97, ed. M.M. Shapiro et al., Kluwer Acad. Publ., Holland.

Simon, M. Heinrich, W., and Mathis, K.D. 1986 Ap. J. 300, 32.

Simpson, J.A. 1983 Ann. Rev. Nucl. Part. Sci. 33, 323.

Swordy, S.P., Muller, D., Meyer, P., L'Heureux, J. and Grunsfeld, J.M. 1990 Ap.J. 349, 625.

Wandel, A., Eichler, D., Letaw, J.R., Silberberg, R. and Tsao, C.H. 1987 Ap.J. 316, 676.

Wandel, A. 1990 21st Internat. Cosmic Ray Conf. (Adelaide) 3, 357.

Webber, W.R. 1981 17th Internat. Cosmic Ray Conf. (Paris), 2, 80.

Webber, W.R., Kish, J.C. and Schrier, D.A. 1990 Phys. Rev. 41, 520,

533, 547, and 566.

Webber, W.R., Kish, J.C. and Schrier, D.A. 1985 19th Interna. Cosmic
  ray Conf. (Lo Jolla) 2, 88.

Wiedenbeck, M.E. and Greiner, D.E. 1980 Ap.J. Lett. 239, L139.

Wiedenbeck, M.E. 1983 18th Internat. Cosmic Ray Conf. (Bangalore) 9,
  147.

Wiedenbeck, M.E. 1985 19th Internat. Cosmic Ray Conf. (La Jolla) 2, 84.

CONTINUOUS REACCELERATION OF COSMIC RAYS IN GALAXIES

M.POHL
Max-Planck-Institut für Radioastronomie
Auf dem Hügel 69
W-5300 Bonn 1
FRG

ABSTRACT. Three physical models for the observed strong turn-over in the total radio spectrum of M33 are discussed. While free-free absorption and the dynamical halo model including ionization and Coulomb losses can be ruled out on the basis of unrealistic implied parameter values, we propose continuous reacceleration by the second order Fermi process as a viable alternative.

1. Introduction

Following the first static models for the propagation of relativistic electrons in the haloes of spiral galaxies later theories for so-called dynamical haloes have predicted spectral breaks by $\Delta\alpha = 0.5$ in the total radio spectra of galaxies, which arise from the competition of different energy loss processes (e.g. Lerche and Schlickeiser, 1982). In a statistical analysis of the total spectra of spiral galaxies Pohl et al. (1991a) have reported that six out of ten tested objects exhibit such breaks centered on frequencies between 0.4GHz and 2GHz. These breaks are smooth and extend over more than two decades in frequency. However, the only galaxy for which a break in the radio spectrum is an eye-catching feature, viz. NGC 4631, revealed a spectral turn-over which was too rapid to be explained by common models.

By including ionization and Coulomb losses Pohl et al. (1991b) have extended the dynamical halo model for electrons of low energies. These additional loss processes result in a second spectral break by $\Delta\alpha = 0.5$ which may merge with the first one to form an effective break by $\Delta\alpha = 1.0$. Applying their model to NGC 4631 these authors report the necessity of gas densities and magnetic field strengths about one order of magnitude larger than expected from earlier investigations.

As an alternative for propagation effects of cosmic ray electrons Israel and Mahoney (1990) have proposed a model, in which the low frequency synchrotron radiation is thought to be absorbed by free-free-transitions in a thermal plasma. A comparison with optically thin free-free-emission at ten GHz or more then gives the constraints to determine the properties of the absorbing plasma. Applying this model to NGC 4631 implies similar excessive gas densities as the Ionization losses model.

In a recent preprint Israel et al. (1992) have combined new data for M33 with

17

M. M. Shapiro et al. (eds.), Particle Astrophysics and Cosmology, 17–22.

data taken from the literature and reported a strong turn-over in the integrated flux density spectrum. Basing on this spectrum it is the aim of this talk firstly to show that applying the two concurring models in the literature (absorption and the extended propagation model) to the integrated radio spectrum of M33 yields implications that are worse than in the case of NGC 4631 and therefore can be ruled out. Secondly, we present an alternative model, which may explain the observed behaviour without unrealistic assumptions on the state of the interstellar medium in this galaxy: the influence of continuous reacceleration on the equilibrium spectrum of electrons.

Due to its rather low influence on the result of the fitting procedure the increasing contribution of optically thin thermal emission to the total intensity at high frequencies has been neglected throughout the paper (see Pohl et al., 1991a).

2. The total radio spectrum of M33

2.1 COMPARISON WITH ABSORPTION MODELS

In the absorption model by Israel and Mahoney it is assumed that synchrotron radiation is emitted with a power-law spectrum, but that almost all of the synchrotron emitting volume is filled by an efficiently absorbing thermal plasma. We base our discussion on the standard solutions of this radiation transport problem, which are given in Israel and Mahoney (1990) for some model geometries. Here we use the homogeneous cylindrical model with a fraction a (=95%) of the height filled with ionized gas. To account for a possible clumpy structure of the ISM in M33 we also discuss the case of small (cylinder-shaped) absorption regions distributed throughout the galaxy.

Fitting of a model spectrum including absorption effects to the observed spectrum of M33 can be done with a statistical significance of nearly 50% for $\tau(327\ MHz) = 1.7$ (for a comparison of the models with the observed spectrum see Fig.1). The optical depth and the (optically thin) thermal flux of a source can been calculated via (Mezger and Henderson, 1967)

$$\tau(\nu) = 0.083\, T_e^{-1.35}\, EM\, \nu^{-2.1}, \qquad S_T = 0.143\, T_e^{-0.35}\, EM\, \Theta^2\, \nu^{-0.1} \quad \text{mJy} \qquad (1)$$

where T_e is in K, EM is the emission measure $\int dl\ n_e^2$ in $pc\,cm^{-6}$, ν is in GHz, and Θ denotes the apparent size in arcmin. By comparison with the observed thermal (free-free) emission we can determine the parameters T_e and EM. The 10.7 GHz observation gives an upper limit for the thermal flux (< 550 mJy).

Israel and Mahoney (1990) have argued for a clumpy structure of the absorbing medium in order to decrease the ratio of thermal emission to the optical depth at a given frequency independently of the parameters T_e and EM. Nevertheless, the following problems remain:

— The synchrotron emission must originate in the same volume as absorption and thus must itself be clumpy.

— Basic radiation transport in a clumpy medium leads to a reduction of EM first (see Pohl et al., 1991b), which does not change the ratio of thermal emission to absorption. Only if the volume filling factor of the absorbing (and synchrotron emitting) regions is so low, that a certain line-of-sight catches statistically less than one clump (surface filling factor $f_A < 1$), the apparent surface can be reduced. As effective emission measure EM we then have to insert the emission measure of a single clump, i.e. the effective line-of-sight is not longer than the typical diameter of one plasma clump.

Estimating $\Theta \approx 30'$ from the radio maps of Buczilowski and Beck (1987) we therefore obtain $T_e \approx 3$ K for $f_A > 1$, which is surely beyond of what can be accepted in rough consistency with the general knowledge on this galaxy, which is well observed for its vicinity to our Galaxy.

2.2 COMPARISON WITH PROPAGATION MODELS INCLUDING IONIZATION LOSSES

A detailed description of the propagation model including ionization and Coulomb losses can be found in Sec.3 of Pohl et al. (1991b). This model yields roughly the same statistical significance as the absorption model. The best fit is obtained with the two breaks melted to one at 1.4 GHz. Since the dependence of the energy loss processes (and thus the corresponding spectral breaks) on the parameters of the environment like the gas density or the magnetic field strength is known, we can calculate these values from the best fit break frequencies and obtain $H_\perp \approx 650$ μG and $n_H \approx 600$ cm^{-3}, where we have neglected the weak dependences on the degree of ionization.

Again, these values are by far too high to be explained in a reasonable scenario for spiral galaxies. They clearly show, that neither the absorption model nor the cooling-flow model (Pohl et al., 1991b) nor any standard leaky box model are appropriate as a general explanation of the observed behaviour, and emphasize the need for a new theoretical interpretation of the unusual radio spectrum from M33.

3. Reacceleration of electrons

In all the above considerations we have neglected any reacceleration process. It will be shown in this section that equilibrium particle spectra modulated by continuous second order Fermi-acceleration can provide a plausible explanation for the observed radio spectrum.

Shock (first order Fermi) acceleration in localized sources is generally thought to be one of the dominant particle acceleration processes. It leads to power-law injection rates with spectral indices not flatter than s=2. This power-law injection $Q(\gamma) = q_0 \gamma^{-s}$ (where γ denotes the particles Lorentz factor) has been the basis for all the propagation models.

On galactic scales an additional particle acceleration process is resonant cyclotron damping of interstellar Alfvén waves with non-degenerate cross helicity state (Schlickeiser, 1989). This interaction of energetic particles with forward and backward running waves can be described by a diffusion term in momentum on the particles distribution function, the so-called momentum diffusion or second order Fermi acceleration. Adding this term to the balance equation (Pohl and Schlickeiser, 1990; see also Schlickeiser et al. 1987) for the electron phase space density f we find

$$\frac{\partial}{\partial \gamma}\left[\gamma^4 \frac{\partial f}{\partial \gamma} + a\gamma^3 f + \eta \gamma^4 f\right] - \lambda \gamma^2 f = -\tau_{F2}\, q_0\, \gamma^{-s} \qquad (2)$$

where the parameters a, η and λ are defined by ratios of time scales at Lorentz factor $\gamma = 1$ for the processes escape (τ_{esc}), radiative losses (τ_{rad}), adiabatic and bremsstrahlung losses (τ_{ad}) and second order Fermi acceleration (τ_{F2}).

$$a = \frac{\tau_{F2}}{\tau_{ad}}, \qquad \eta = \frac{\tau_{F2}}{\tau_{rad}}, \qquad \lambda = \frac{\tau_{F2}}{\tau_{esc}}.$$

This model is a general expansion of the common dynamical halo models, which take into consideration the interplay of adiabatic losses (here with parameter a) and radiative energy losses (here with parameter η). Therefore, if $a \gg 1$, Eq.5 describes the standard dynamical halo revealing itself by a spectral break of $\Delta\alpha = 0.5$ in the radio spectrum at the critical electron energy $\gamma_{c1} = a\eta^{-1}$. In this section we want to discuss the case $a < 1$, i.e. efficient reacceleration of cosmic rays.

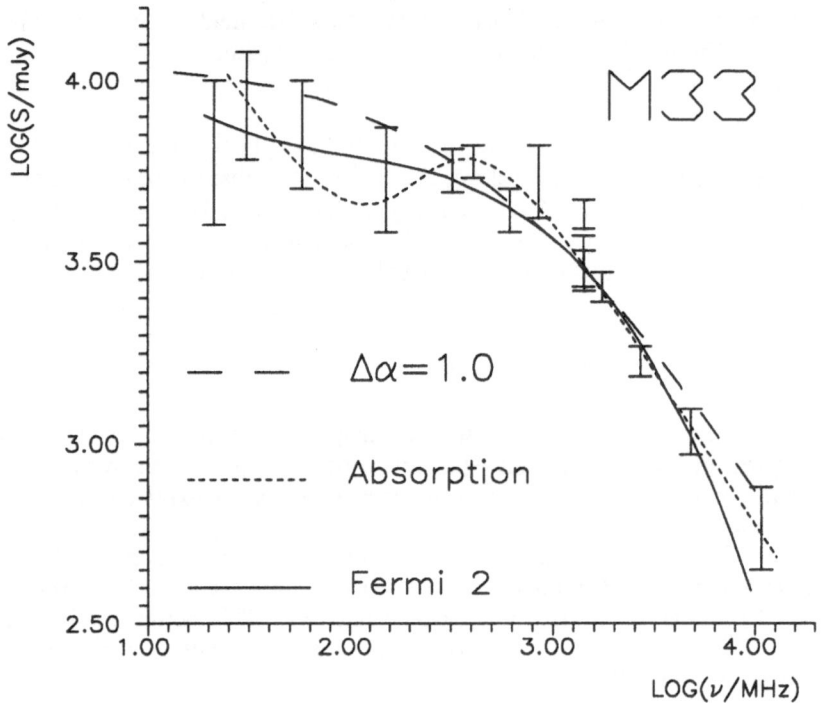

Fig.1: *The integrated radio spectrum of M33 as taken from Israel et al. (1992). We have included the best fits of the concurring models, the absorption model, the extended propagation model including Ionization and Coulomb losses and the propagation model including continuous reacceleration by the second order Fermi process.*

The Green's function solution to this equation is given in Schlickeiser (1984). The solution for power-law injection can be found in Crusius (1988). For ease of exposition ionization losses have been neglected here.

Considering asymptotic particle spectra for long escape time scales ($\lambda < 1$), the low energy ($\gamma < \eta^{-1}$) number density of electrons $N(\gamma) = 4\pi\,\gamma^2\,f(\gamma)$ changes from what is predicted by the common propagation models, if $a < 1$, i.e. the time scale for second order Fermi acceleration is shorter than that for adiabatic cooling and non-thermal bremsstrahlung. Depending on the quality of confinement (λ) the energy spectral index is $1 + \lambda$ (≈ 1 for $\lambda \to 0$). At high energies we obtain the old result of the injected power-law steepened by 1 together with a bump at an energy slightly above the characteristic one $\gamma_c = \eta^{-1}$.

We have fitted a synchrotron spectrum corresponding to the above described model, which agrees with the observations with about the same significance (50%) as the fits of the concurring models. The best fit spectrum (for $\lambda = 0.4$) is shown together with the observed integrated flux density spectrum and the other models in Fig.1. A slight variation of the critical energy γ_c over the galactic volume leads to $\lambda < 0.4$ in the best fit.

The main assumption used in this model is that the time scale for second order Fermi acceleration is shorter than the time scale for adiabatic and bremsstrahlung losses.

4. Summary and conclusions

We have rediscussed two concurring models in the literature, which explain strong turn-overs in the total radio spectra of galaxies, on the basis of data on M33. Both free-free absorption and the propagation model including ionization losses were found to imply parameters which are orders of magnitude from realistic ones, worse than those required to explain the spectrum of NGC 4631.

We have proposed continuous reacceleration by the second order Fermi process to be an alternative explanation for the observed behaviour, which does not demand drastic assumptions on the state of the interstellar medium.

Thus the following conclusions can be drawn:

1) In case of M33 the cooling flow scenarios of the propagation models including ionization losses are much harder to legitimate as for NGC 4631. Also absorption models can be ruled out for reproductions of the observed spectrum. Although absorption or ionization losses must occur for physical reasons, they can not influence the total radio spectra of spiral galaxies at frequencies above 0.3 GHz.

2) Second order Fermi acceleration is important as modulator of the low energy particle spectra of cosmic rays. Momentum diffusion provides a natural explanation for the strong flattening of the total radio spectra towards lower frequencies in some galaxies (M33, NGC 4631).

3) The hypothesis of momentum diffusion can be in principle tested by investigations of local radio spectra. Thus Eq.2 has to be extended in order to describe spatial transport like diffusion and convection.

4) It is rather likely that many more than the two galaxies mentioned in this talk show strong turn-overs in their integrated flux density spectrum. Therefore, more low frequency radio observations are needed to obtain a reliable data basis for a similar analysis of the spectra of other galaxies.

References

Buczilowski, U.R., Beck, R. (1987) 'A multifrequency radio continuum survey of M33. I. Observations', *AAS 68*, 171-185

Crusius, A. (1988) 'Plasma phenomena in synchrotron theory and the consequences for astrophysics', *PhD-Thesis*, University of Bonn, Germany

Israel, F.P., Mahoney, M.J. (1990) 'Low-frequency radio continuum evidence for cool ionized gas in normal spiral galaxies', *ApJ 352*, 30-43

Israel, F.P., Mahoney, M.J., Howarth, N. (1992) 'The integrated radio continuum spectrum of M33: evidence for free-free absorption by cool ionized gas', *AA*, in press

Lerche, I., Schlickeiser, R. (1982) 'On the transport and propagation of relativistic electrons in galaxies. The effect of adiabatic deceleration in a galactic wind for the steady-state case', *AA 107*, 148-160

Mezger, P.G., Henderson, A.P. (1967) 'Galactic HII regions. I. Observations of their continuum radiation at the frequency 5 GHz', *ApJ 147*, 471-489

Pohl, M., Schlickeiser, R. (1990) 'The influence of extended source distributions on cosmic ray spectral index variations in the galactic wind model', *AA 234*, 147-155

Pohl, M., Schlickeiser, R., Hummel, E. (1991a) 'A statistical analysis of cosmic ray propagation effects on the low frequency radio spectrum of spiral galaxies', *AA 250*, 302-311

Pohl, M., Schlickeiser, R., Lesch, H. (1991b) 'Imprints of a cooling flow on the radio continuum spectrum of NGC 4631', *AA 252*, 493-497

Schlickeiser, R. (1984) 'An explanation for abrupt cutoffs in the optical-infrared spectra of non-thermal sources. A new pile-up mechanism for relativistic electron spectra', *AA 136*, 227-236

Schlickeiser, R. (1989) 'Cosmic ray transport and propagation. I. Derivation of the kinetic equation and application to cosmic rays in static cold media', *ApJ 336*, 243-263

Schlickeiser, R., Sievers, A., Thiemann, H. (1987) 'The diffuse radio emission from the Coma cluster', *AA 182*, 21-35

COSMIC RAY COMPOSITION AT AND ABOVE THE KNEE

Todor Stanev
Bartol Research Institute
University of Delaware
Newark, DE 19716, U.S.A.

ABSTRACT. Recent data shows that the cosmic ray composition changes rapidly in the most interesting regions of the cosmic ray spectrum. The fraction of heavy nuclei grows strongly when approaching the 'knee' and decreases significantly before the 'ankle' of the spectrum. While such behaviour is not unexpected on the basis of cosmic ray acceleration and propagation models, the new results give meaningful limits on the variety of models of the cosmic ray origin.

1. INTRODUCTION

The cosmic ray spectrum extends to about 10^{11} GeV, i.e. over 11 decades in total energy per particle. It is quite obvious, and it has been shown in numerous calculations, that no single acceleration mechanism can support particle acceleration over such range. At highly ordered man made particle accelerators, the acceleration to about 10^3 GeV proceeds in numerous injection and pre-acceleration stages. The fact that the cosmic ray spectrum extends to such high energy defies trivial explanations and implies a number of natural acceleration processes.

Fig. 1 shows data on the cosmic ray spectrum above 10^3 GeV/nucleus with the flux multiplied by $E^{2.75}$. The solid points represent direct measurements, while the open ones refer to measurements derived from air shower data. Because the air shower measurements estimate the total amount of energy carried by a cosmic ray nucleus we have chosen to discuss the composition in the appropriate terms of energy/nucleus. The data set includes two regions of special interest in the spectrum: the *knee* region, where the slope of the spectrum increases by $\Delta_\gamma = 0.3\text{-}0.4$, and the *ankle* where the spectrum might be flattening again.

A *standard model* describing the origin of cosmic rays has evolved over the past 15 years (Blandford & Eichler, 1987; Jones & Ellison, 1991), and accounts naturally for the major observed features of the cosmic ray spectrum, *viz.*, the energy density of cosmic ray nuclei, the slope of the power law energy spectrum, and the increase in the ratio of primary to secondary nuclei with energy. In the *standard model*, the power source is thought to be

M. M. Shapiro et al. (eds.), Particle Astrophysics and Cosmology, 23–34.
© 1993 *Kluwer Academic Publishers*.

supernova explosions (Ginzburg & Syrovatskii, 1964). Diffusive acceleration at supernova blast waves (Axford, Leer & Skadron, 1977; Krimsky, 1977; Bell, 1978; Blandford & Ostriker, 1978), together with rigidity-dependent escape from the galaxy (Berezinsky *et al.* 1990), account for the value of the spectral index as well as the energy-dependence of the secondary to primary nuclei ratio. The *standard model* works very well in the GeV energy range, in the very beginning of Fig. 1, but fails at higher energy.

The reason is the difficulty of achieving a maximum energy per particle which is high enough to account for the full cosmic ray spectrum. Lagage & Cesarsky (1983) showed that the maximum energy per particle from a typical supernova ($\sim M_\odot$ ejected with kinetic energy $\sim 10^{51}$ erg into the interstellar medium with $B \sim 3\mu$Gauss) is $E^{max} \sim Z \times 100$ TeV, where Z is the particle's charge. This maximum energy can be extended in various ways or a new acceleration mechanism can take over and produce the observed cosmic rays of higher energy. Several such possibilities have been discussed in the literature.

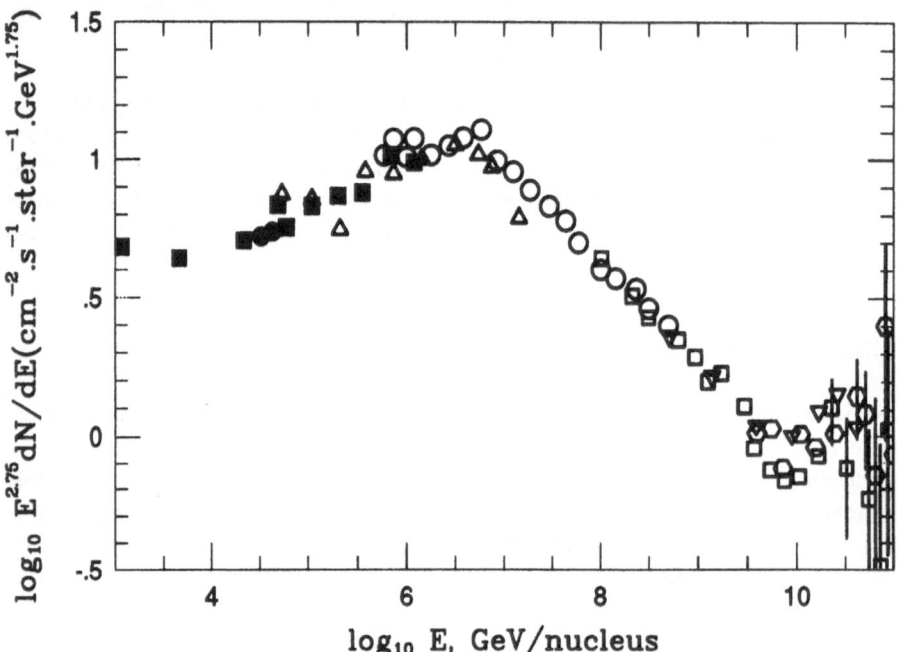

Fig. 1. The cosmic ray spectrum in terms of E/nucleus. The full symbols are from Grigirov *et al.* (1971) and JACEE (private communication). The open symbols are from air shower experiments (Danilova *et al.* 1977; Nagano *et al.* 1984; and Cassiday *et al.* 1990).

In this talk we shall argue that different types of processes affect the composition of the cosmic rays in unique ways and thus the composition holds the key to understanding the origin of high energy cosmic rays. We shall start with a brief review of the composition measurements, discuss two types of acceleration models developed to explain the spectrum in the region of the *knee* and the associated composition. Finally we shall review the experimental data at the higher end of the cosmic ray spectrum, the recent results of the Fly's Eye experiment on the composition at 10^{18} eV, and relate them to the common belief that the highest energy cosmic rays are of extragalactic origin.

2. COMPOSITION ABOVE 1 TeV.

There are two quite different methods for studying cosmic ray spectra and composition above 1 TeV/nucleus. Direct experiments have now extended the measured energy region up to 100 TeV/nucleus. At higher energies the cosmic ray flux becomes extremely low, less than one particle per m^2.s.sr and the current statistics of direct experiments is insufficient for even an analysis of the spectrum, not to mention composition. The flux continues decreasing at the rate of 1/50 per energy decade and extensive air showers with an area greater than 10^4 m^2 become the only source of information. The determination of cosmic ray composition from air showers is however very uncertain and since the conversion from measured quantities to primary energy depends strongly on the mass of the primary particle, so is the primary spectrum.

2.1. Direct measurements.

Since primary cosmic rays interact in the first tenth of the atmosphere direct measurements of the primary composition involve lifting an instrument to an altitude where the amount of intervening matter is negligible, i.e. use either a satellite of high altitude balloons. Three different direct experiments have measured the composition above 1 TeV/nucleus in this way. HEAO 3-2C (Engelmann et al. 1990) is a satellite experiment, that has detected nuclei heavier than beryllium up to ~few TeV/nucleus. CRN (Swordy et al. 1990) flew for a week on the Space Shuttle and detected nuclei heavier than carbon in the region 1 to 100 TeV/nucleus. JACEE (Asakimori et al. 1991a) measured all components of the cosmic ray flux in a series of baloon flights and has a limited statistics above 100 TeV/nucleus. The estimate of the primary energy of individual events seen by JACEE depends on the fraction of the energy converted to γ-rays, K_γ. Apart from systematic uncertainties K_γ is subject to large fluctuations, especially for protons, and the energy estimate is accurate only for relatively large samples.

Fig. 2 shows the fluxes of oxygen and iron detected by all three experiments. The absolute values of the detected fluxes agree very well in the regions where different experiments overlap. It should be noted that up to energies around 1 TeV/nucleus, iron nuclei may be subjected to solar modulation. At somewhat higher energies, both types of nuclei show spectra flatter than the canonical spectral slope $\gamma=2.75$ as well as an almost identical spectral flattening with increasing energy. Up to a few tens of TeV/nucleus, the spectra could be explained (Engelmann et al. 1990; Swordy et al. 1990) by a simple leaky box model with a source spectrum having $\gamma=2.2\pm0.1$ and a rigidity dependent escape length of $1/\lambda \sim R^{-0.6}$. Such a model predicts a steepening of the spectra of the individual components at higher energy, which is not evident from the JACEE data. Because of the large error bars on the JACEE data points, the contradiction is not very significant statistically, although all groups of nuclei show a similar trend. Detailed comparisons for other nuclei are not easy because of the different selection criteria used in the experiments.

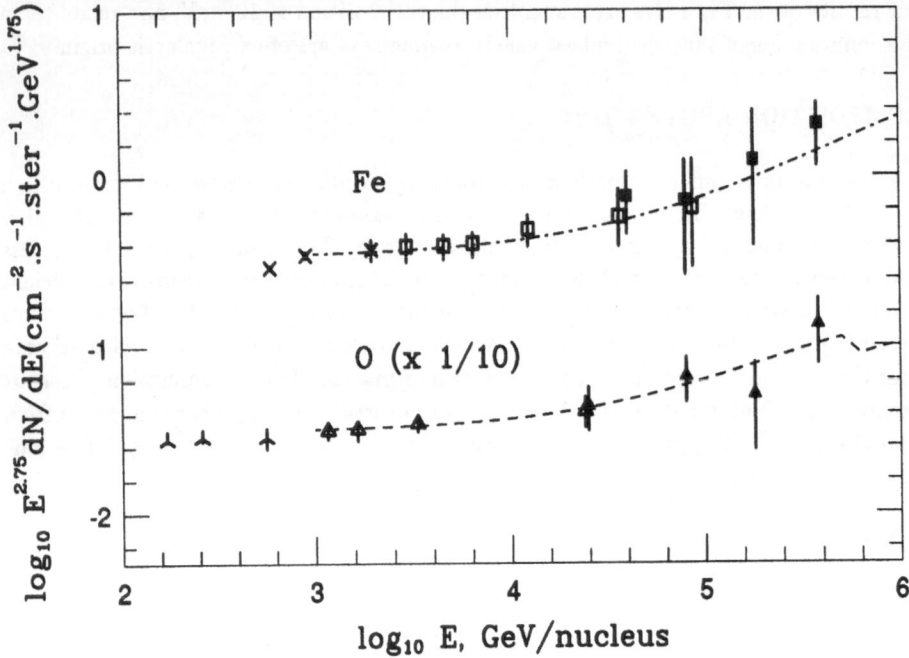

Fig. 2. Iron and oxygen spectra from direct measurements. The data is in order of increasing energy from HEAO-3C (Engelmann *et al.* 1990), CRN (Swordy *et al.* 1990) and JACEE (Asakimori *et al.* 1991a). One half of the sum of carbon and oxygen spectra from JACEE is plotted. The oxygen spectrum is divided by 10. Lines are from the model of Fig. 5.

Hydrogen and helium spectra in this energy range have only been measured by JACEE (Asakimori *et al.* 1991b). Fig. 3 shows the measured spectra of these two components. The last two points of the proton spectrum, which contain altogether eight events, indicate a sharp steepening of the proton spectrum. The data point at ∼133 TeV is about 3σ below the best fit for all points. Compared to the central value of the expectations from a fit based on data between 6 and 40 TeV , however, this point is lower by more than 8σ. Even though the statistics is low, the indication for a cut-off of the proton spectrum is quite strong. There are no indications in the JACEE data for a similar cutoff in the He spectrum, which is detected by this experiment as the highest component of the cosmic ray flux.

JACEE has limited statistics with only 51 nuclei with energy above 100 TeV/nucleus, which corresponds to 0.25 particles.$(m^2.hr.sr)^{-1}$ and an energy flux of 110 ergs.$(m^2.hr.sr)^{-1}$. As can be seen from Fig. 1, the JACEE points in the full particle spectrum (full circles) are in good quantitative agreement with the satellite data from the Proton series (Grigorov *et al.* 1971) and derivations from air shower experiments. Only 6 of these 51 events (0.034 $(m^2.hr.sr)^{-1}$) and 12 ergs.$(m^2.hr.sr)^{-1}$ are protons.

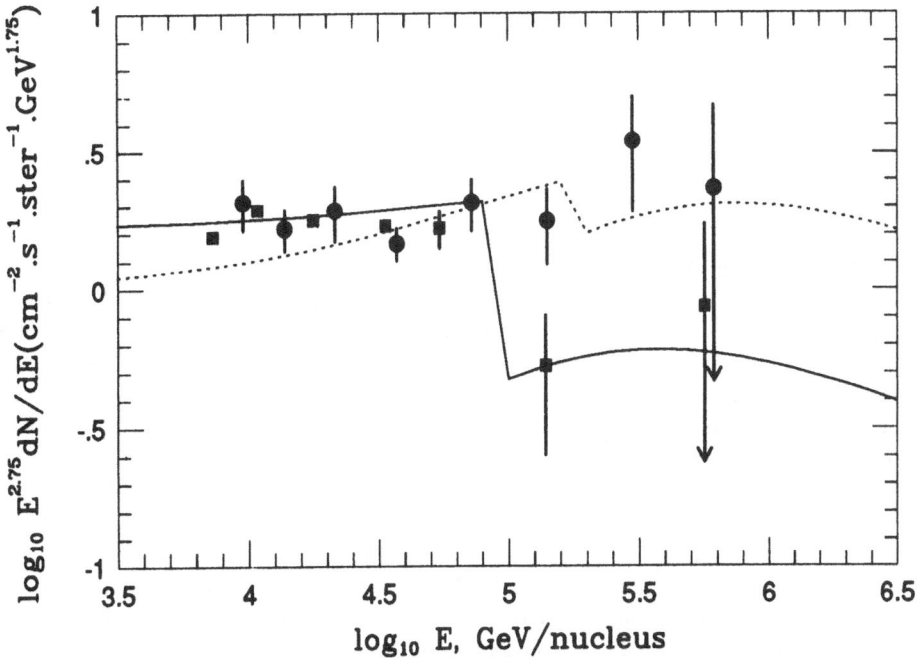

Fig. 3. Proton (squares) and He (circles) spectra from JACEE (Asakimori *et al.* 1991b). Lines are from the model of Fig. 5.

2.2 Air Shower Measurements.

The factor of $E^{2.75}$ in Fig. 1 was chosen to emphasize the main feature of the cosmic ray spectrum—the change in slope between 10^3 and 10^4 TeV, in the region identified as the "knee". Data also shows a flattening of the spectrum just before the "knee". The data points in this region come from air shower experiments that can only estimate statistically the energy of the primary cosmic rays from the measured characteristics of the extensive air showers that they initiate in the atmosphere.

The sensitivity to the mass of the primary comes from the different rate of energy dissipation in the atmosphere. Since nuclear cross-sections are significantly larger than the proton inelastic cross-section, showers initiated by heavy nuclei start higher in the atmosphere. Nuclei also disintegrate into their constituent nucleons thus speeding up the shower development and causing faster absorption in the atmosphere. The actual measurements are made at Earth's surface, at mountain altitude at best, where air showers initiated by nuclei of different mass have, in principle, observable differences. Though the mass resolution cannot be very good, shower experiments are expected to distinguish between the four or five groups of primary nuclei thought to represent the major components of the cosmic ray flux.

There are generally two types of shower experiments. The first one measures the bulk of the shower particles, mostly electrons and photons which result from many generations of secondary hadron interactions and subsequent electromagnetic cascades. Important quantities for this type of measurements, which are sensitive to the primary mass, are the rate of absorption of the showers in the atmosphere and the ratio of the numbers of all charged particles N_e and GeV muons N_μ at the observation level.

The second type of experiments concentrate on the measurement of high energy remnants of the first interaction of the primary nucleus in the atmosphere. These include the detailed investigation of the structure of the hadronic core of the air shower, which is done with thick electronic or emulsion calorimeters, and the observation of TeV muon groups in deep underground detectors.

Since the shower development is a fairly complicated process that involves multiple hadronic and electromagnetic branches, the relation between observable shower characteristics and the mass and energy of the primary nuclei is provided by theoretical shower development models. Most of the the current models use montecarlo technique to follow the interactions of shower hadrons and analytic extensions for the description of the low energy electromagnetic cascades.

Shower development is sensitive primarily to the forward fragmentation region of the inelastic hadronic interactions, which is studied at accelerators in the restricted fixed target energy range. Thus, part of the the ambiguity in the reconstruction of the type and energy of the primary results from uncertainties in extending the high energy particle physics to the air shower energy range. Although different experiments observe qualitatively the same effects, they might reach different conclusions, depending on the underlying particle physics model.

Using collections of air shower data, different groups often reach contradictory conclusions. While some authors see the emergence of a new component, consisting almost purely of hydrogen nuclei (Fichtel & Linsley, 1986) others claim that heavy nuclei dominate (Freudenreich *et al.* 1990). Since the derivation of the primary particle energy depends quite strongly on the estimated (or assumed) mass, uncertainty in composition leads to a corresponding uncertainty in the energy spectrum. Although the change of the slope is well established, we cannot claim to know the exact shape of the cosmic ray spectrum at the "knee".

3. ACCELERATION MODELS.

It is generally thought that galactic cosmic rays with energies up to about $10^{14}eV/n$ are accelerated at shock waves associated with supernova remnants (SNR's) (e.g., Völk, Zank & Zank (1988), Drury (1992)). In this picture, the very energetic blast wave propagates into the tenuous interstellar medium (ISM), heating and compressing the gas. A fraction of the inflowing thermal gas becomes sufficiently energized that it can scatter off magnetic irregularities. The scattered particles can then random walk back and forth across the shock. The particle momentum increases with each cycle and an upper limit on the particle energy gain is determined by the finite lifetime of the shock and the particle acceleration time scale. We already referred to the 100 TeV × Z limit of Lagage & Cesarsky. Völk, Zank & Zank (1988) have used a time dependent onion shell supernova model including various ISM magnetic field values and different expansion modes to simulate the acceleration process. Apart from small differences in the spectrum shape, the results is still the same: SNR's are able to producing the interstellar cosmic ray population up to about 10^{14}-$10^{15} eV/n$, at which point the spectrum ends abruptly. Unless a mechanism quite different from that of diffusive shock acceleration is operative at SNR's, it is difficult to see how SNR's can generate the very highest energy cosmic rays beyond $10^{15} eV/n$. Nonetheless, this result provides us with simple means for tentatively identifying the nature of the mechanism or process responsible for accelerating the highest energy cosmic ray component, provided our knowledge of the composition data can be extended up to about $10^{15} eV$.

Consider two of the suggestions for extending cosmic ray energies up to $10^{18} eV$. The first, advocated by Ip & Axford (1992), involves the reacceleration of cosmic rays on larger scales, possibly in superbubbles resulting from the interaction of multiple supernova remnants and stellar winds. As shown by Ip & Axford, such a reacceleration process results in a gradual enrichment of heavy nuclei relative to protons and helium, an effect which is similar to rigidity-dependent propagation. Fig. 4, which shows our representation of their model, illustrates this behaviour.

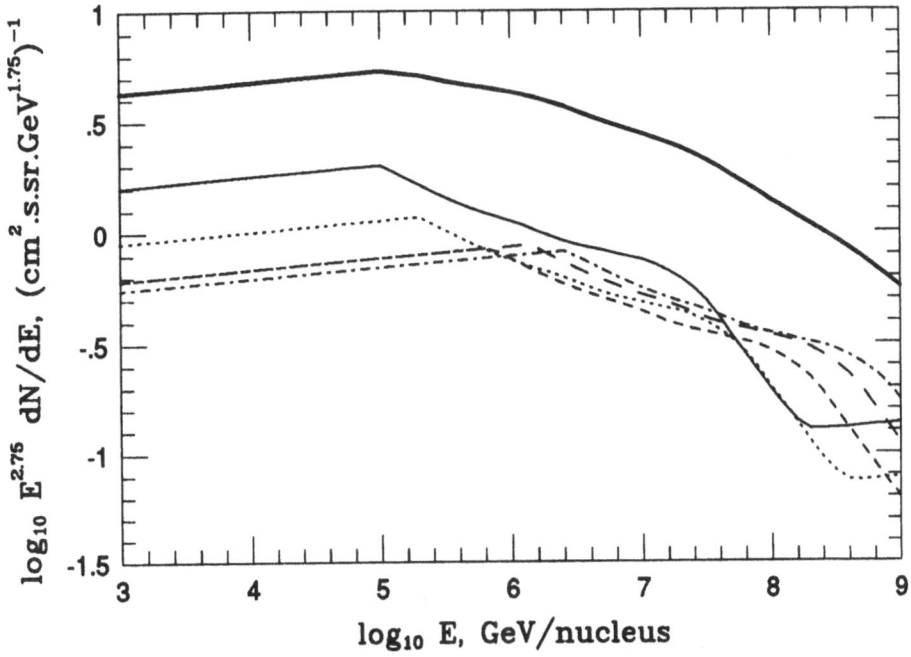

Fig. 4. Cosmic ray spectrum and composition from Ip&Axford. The thin lines are for different components: H (full line), He (dots), CNO (short dashes), Si-Mg (long dashes), Fe (dash-dot). The thick line is the sum - the all particle spectrum.

A second class of models involves a transition either to a new accelerator (Wdowczyk & Wolfendale, 1989; Protheroe, 1992) or to drastically different physical conditions in the interstellar medium surrounding the supernova remnant (Völk & Biermann (1988); Silberberg et al. 1990). The sharp cut-off in energy resulting from the "standard" SNR acceleration model will produce an abrupt change in spectral composition, from a spectrum dominated by protons and helium to one dominated by heavy nuclei. Thereafter, a transition to a spectral composition characteristic of the new accelerator should appear. The first transition is therefore the sharp rigidity-dependent cutoff of the supernova blast wave accelerator.

To exhibit the composition that can arise from an independent accelerator class of models, we have constructed a phenomenological model of the cosmic ray spectrum which includes two components. The first is due to a "standard" SNR shock model and is characterized by the observed low energy composition and an abrupt rigidity cut-off at 90 TeV. The second component begins with a flat ($\gamma=2$) differential spectrum and has a smooth transition to the observed $\gamma=3.1$ spectrum at a rigidity of 200 TeV. The shape and composition of the second component are tuned to reproduce the spectral features and the energy dependence of the heavier components shown in Figs. 2 and 3. The resulting composition and particle flux are shown on Fig. 5.

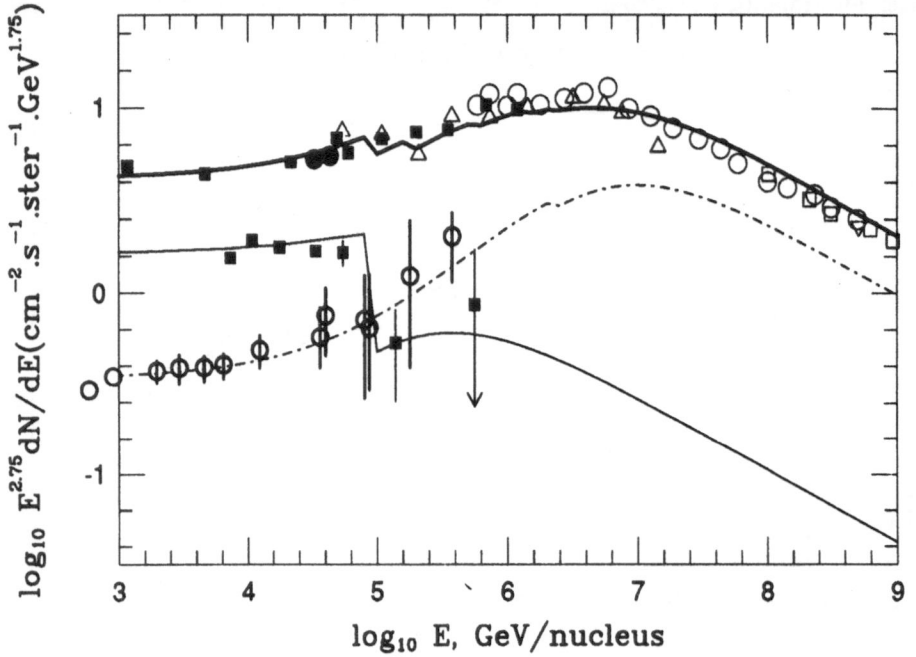

Fig. 5. Two component phenomenological model of the cosmic ray spectrum and composition compared to the data set for the all particle flux and the H and Fe components. The line codes are the same as in Fig. 4.

The transition to the region whose composition is determined fully by the second acceleration mechanism on Fig. 5 happens only at an energy $\sim 10^{16}$ eV, and at even higher energies in the model of Ip & Axford (1992). The characteristic changes for each class of models can, however, be observed at energies lower than those at the knee. By extending the composition data to 10^{15} eV one could detect the differences between the two classes of models quite easily. Whereas the continuous reacceleration model of Ip

& Axford predicts that protons represent 34% of the total flux at energies between 10^{14} and 10^{15} eV, the second source model of Fig. 5 predicts only 8.5%. The corresponding fraction of iron nuclei is very similar (14% and 15% respectively), but the effective energy spectra of the iron component are very different having spectral indices $\gamma=2.7$ and $\gamma=2.3$. These differences are certainly measurable by direct experiments if the current statistics is increased by a factor of 10. An additional virtue of such measurement will be the overlap with the EAS energy region, which will allow a renormalization of the cascade development models and will significantly improve the quality of the air shower data analysis.

4. THE COMPOSITION AT THE ANKLE OF THE SPECTRUM.

An attempt was made recently (Gaisser *et al.*, 1992) to use the data of the Fly's Eye experiment (Cassiday *et al.*, 1990) to study the composition in the vicinity of 10^{18} eV. In a general way the Fly's Eye technique fits the definition of the type I shower experiment, i.e. it detects the bulk of shower particles. There is one major difference, though, with the conventional shower array. The Fly's Eye detects the flourescent light induced by the charged particles in the cascade, which only depends weakly on the atmospheric density. Thus the experiment is in principle able to map the shower longitudinal development and use the integral amount of flourescent light for an estimate of the primary energy. The sensitivity to composition comes from the shape of the shower profile and is usually expressed in terms of the depth of the shower maximum X_{max}.

The Fly's Eye has recently increased its statistics of stereo showers (showers observed simultaneously with FE-I and FE-II, located about two miles apart) of energy above 3×10^{17} eV to more than 2000 showers. Massive computer simulations were performed for showers induced by nuclei with different primary mass and three different particle physics models. The simulations included the experimental triggering conditions and analysis algorithms, i.e. they attempted to account for the experimental biases. The simulated X_{max} distributions from single nuclar components were than used to fit the experimental data sample. The fraction of showers needed from each component to fit the experimental distributions dtermines the presence of this component in the cosmic ray flux. Table I shows the results of the fits for three energy ranges around 10^{18} eV (= 1 EeV).

Table I. Results from the Fly's Eye composition fit.

E, EeV	Number	KNP Model			mini-jet model		
		H	Fe	χ^2	H	Fe	χ^2
0.3–0.5	994	0.21±0.07	0.79±0.11	2.51	0.12±0.03	0.88±0.06	4.85
0.5–1.0	867	0.27±0.12	0.66±0.12	1.56	0.21±0.10	0.79±0.16	3.17
> 1.0	690	0.43±0.04	0.56±0.05	0.96	0.39±0.15	0.61±0.17	1.32

Fig. 6 shows an example for the fit quality in the energy range above 10^{18} eV. Two comments should be made before the discussion of Table I in astrophysical terms:
(i) The fit was performed with three groups of nuclei: H, CNO and Fe. The reason is that

previous calculations have shown that at this energy H and He generated showers are very similar to each other and undistinguishable by the experimental technique. The same is true for Si and Fe showers.

(ii) The fitting procedure allows very little or none of the CNO component in the fit. One of the possible reasons might be that the X_{max} fluctuations of showers initiated by heavy nuclei are currently dominated by reception and analysis biases.

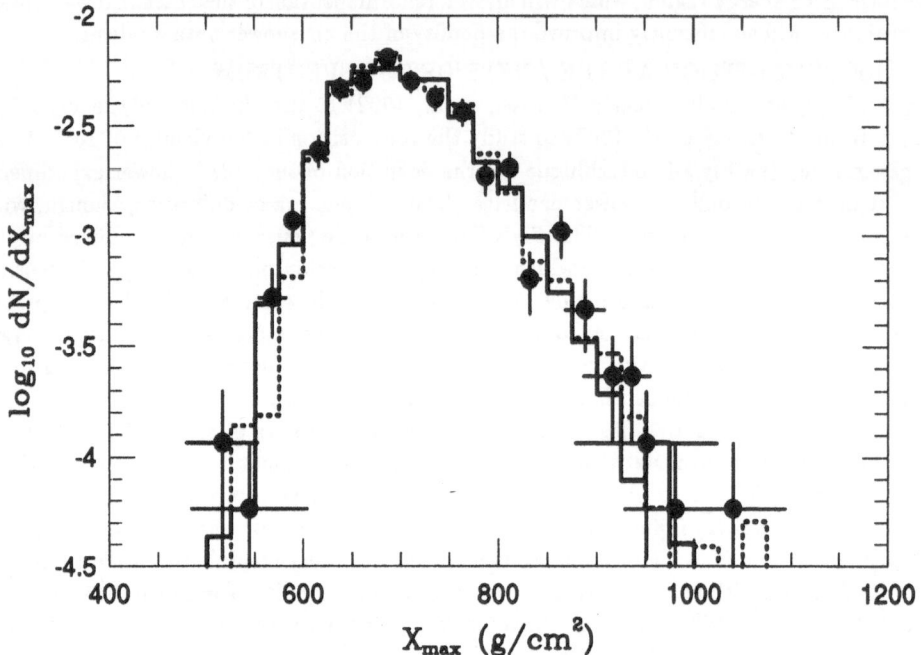

Fig. 6. Comparison of the experimental X_{max} distribution (dots) with the fitted Monte-carlo distribution for E>1 TeV and two interaction models (from Gaisser *et al.* 1992).

Other than that the fits in Table I are fully consistent with the picture one expects from the cosmic ray spectra shown on Figs. 4 and 5. Below 10^{18} eV the cosmic ray flux is dominated by very heavy nuclei. One can speculate that around 10^{18} eV an extragalactic flux becomes detectable and the fraction of of proton generated showers increases. The extragalctic component has to be purely protonic because energetic heavy nuclei shoud be desintegrated in collisions on the $3_\circ$K background. Although I am not ready yet to give numerical values for the emerging extragalactic proton flux (and everybody can do that using the fractions in Table I and his favorite total flux value), it is certainly consistent with at least one of the models (Rachen and Biermann, 1992), based on particle acceleration in AGN jets.

This example shows how powerful the knowledge of the cosmic ray composition is as a tool for studies of the origin of cosmic rays. At 10^{18} eV there is not a change of the spectral index. The composition analysis is able to detect the extragalctic component before it starts dominating the spectrum. An improved statistics and analysis program, which we expect from the upcoming Hi Resolution Fly's Eye and other giand air shower experiments, should be able to determine with better accuracy the magnitude and the slope of the extragalactic component, and pass a solid judgement on its origin and cosmological evolution.

ACKNOWLEDGEMENTS. Most of the work on which this talk is based is done in close collaboration with T.K. Gaisser. The author is gratefull to the JACEE collaboration and the Fly's Eye group for all the information and data that they shared with hime through the years. This work is supported in part by National Science Foundation and NASA. I also wish to thank the organizers of this course, Drs. M.M. Shapiro and J. Wefel, and the Center for Scientific Culture 'E. Majorana', for their hospitality and for the creative atmosphere in Erice.

REFERENCES

Asakimori, K. *et al.* 1991a, in *Proc. 22nd Int. Cosmic Ray Conf, Dublin, Ireland* (Dublin Institute for Advanced Studies, Dublin, Ireland, 1991) **2** 57.

Asakimori, K. *et al.* 1991b, in *Proc. 22nd Int. Cosmic Ray Conf. Dublin, Ireland* (Dublin Institute for Advanced Studies, Dublin, Ireland, 1991) **2** 97.

Axford, W.I., Leer, E. and Skadron, G. 1977, in *Proc. 15th Int. Cosmic Ray Conf. Plovdiv, Bulgaria* (Bulgarian Academy of Sciences, 1977) **11**, 132.

Bell, A.R. 1978, *M.N.R.A.S.* **182**, 147.

Berezinsky, V.S., Bulanov, S,V., Dogiel, V.A., Ginzburg, V.L. and Ptuskin, V.S., *Astrophysics of Cosmic Rays*, ed. by V.L. Ginzburg (North Holland, Amsterdam, 1990).

Blandford, R.D. and Eichler, D., 1987, *Phys. Reports*, **154**, 1.

Blanford, R. D. and Ostriker, J., 1978, *Ap. J.* **221**, L29.

Cassiday, G.L. *et al.* 1990, *Ap. J.* **356**, 669.

Danilova T.V. *et al.* 1977, in *Proc. 15th Int. Cosmic Ray Conf. Plovdiv, Bulgaria* (Bulgarian Academy of Sciences, 1977).

Drury, L.O'C. 1992, in *Particle Acceleration in Cosmic Plasmas*, ed.'s G.P. Zank and T.K. Gaisser, IAP Conference Proceedings 264, p. 189.

Engelmann, J.J. *et al.* 1990, *Astron. Astrophys.*, **233**, 96.

Fichtel, C.E. and Linsley, J. 1986 *Ap. J.* **300**, 474.

Freudenreich, H.T. *et al.* 1990 *Phys. Rev.* **D41** 2732.

Gaisser, T.K. *et al.* 1992, subitted to *Phys. Rev. D.*

Ginzburg, V.L. and Syrovatskii, S.I. *The Origin of Cosmic rays*, (Pergamon Press, 1964).

Grigorov, N.L. *et al.* 1971, in *Proc 12th Int. Cosmic Ray Conf. Hobart, Tasmania* (Univ. of Tasmania Press, Hobart, Tasmania, 1971) **2** 206.

Ip, W.H. and Axford, W.I. 1992, in *Particle Acceleration in Cosmic Plasmas*, ed.'s G.P. Zank and T.K. Gaisser, AIP Conference Proceedings 264, p. 400.

Jones, F.C. and Ellison, D.C., 1991, *Space Sci. Rev.*, **58**, 259.

Krimsky, G.F. 1977, *Dokl. Akad. Nauk SSSR*, **243**, 1306.

Lagage, P.O. and Cesarsky, C.J. 1983, *Astron. Astrophys.* **125** 249.

Nagano, M. *et al.* 1984, *J. Phys G*, **10**, 1295.

Protheroe, R.J. and Szabo, A.P. 1992, submitted to *Nature*.

Rachen, J. and Biermann, P.L. 1992, in *Particle Acceleration in Cosmic Plasmas*, ed.'s G.P. Zank and T.K. Gaisser, AIP Conference Proceedings 264, p. 393.

Silberberg, R., Tsao, S.H., Shapiro, M.M. and Biermann P.L. 1990, *Ap. J.* **363**, 265.

Swordy, S.P. *et al.* 1990, *Ap. J.* **349**, 625.

Völk, H.J., Zank, L.A. and Zank, G.P. 1988, *Astron. Astrophys.* **198**, 274.

Völk, H.J. and Biermann, P.L. 1988, *Ap. J. Lett.* **333**, L65.

Wdowczyk, J.and Wolfendale, A.W. 1989, *Ann. Rev. Nucl. Part. Sci.* **39** 43.

COSMIC RAYS OF THE HIGHEST ENERGIES

X. Chi, J. Wdowczyk and A.W. Wolfendale
Physics Department,
University of Durham,
South Road,
Durham, DH1 3LE, UK.

ABSTRACT. The results of recent studies of ultra-high energy cosmic rays, largely by Chi et al., are summarised. It is argued that the subject is in a most interesting state in that anisotropies are present above 10^{19} eV and hints are appearing as to the nature of the sources. It is argued that the composition is mixed - largely protons and iron - and that some of the particles are of extragalactic origin. It is appreciated that some of the ideas are contentious but it is felt that their chance of being true is sufficient to justify the construction of bigger extensive air shower arrays.

1. Introduction

1.1. THE BASIC PROBLEM

The fact that the majority of very energetic cosmic rays appear to be atomic nuclei, coupled with the presence of a magnetic field in the Galaxy - of very uncertain topography - means that the determination of the origin of the particles is difficult. However, above 10^{19} eV the situation improves for protons and 'local' sources, by 'local' being meant sources within about 1 kpc. It is immediately apparent that there is a great need to a select out particles which are predominantly protons and to look in directions where there are likely to be local sources. This is the thrust in our recent series of papers (Chi et al., 1992 **a, b, c** and **d**).

1.2. THE GALACTIC PLANE ENHANCEMENT FACTOR

Some years ago Wdowczyk and Wolfendale (1984) examined the then EAS data to search for the presence of an asymmetry in arrival directions favouring the Galactic Plane. They defined the intensity as a function of Galactic latitude by

$$I(b) = I_o \left[(1 - f_E) + f_E \cdot a \exp -b^2 \right]$$

where a is a normalisation factor and b is in radians.

M. M. Shapiro et al. (eds.), Particle Astrophysics and Cosmology, 35–42.

36

and determined f_E as a function of assigned energy. Figure 1 shows the latest compendium of values of f_E; it will be noted that the evidence favouring an increase of f_E with energy is quite marked - at least until above 3×10^{19} eV. We regard this result as comprising still the best indication for a Galactic origin of at least some of the particles above 10^{19} eV.

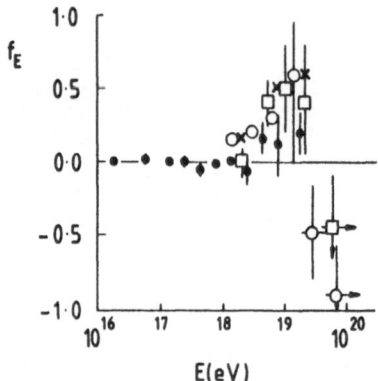

Figure 1 Average Galactic Plane Enhancement factor, f_E, versus energy for the different arrays. The results relate to all longitudes and thus have different longitude weightings, a particularly significant fact when comparison is made between the Sydney results (Southern hemisphere) and the rest.

2. Very Recent Studies of the Problem

2.1 FURTHER ANALYSIS OF THE ENHANCEMENT FACTOR

It was remarked in the previous section that local proton sources might be detectable. It is very likely that cosmic ray sources would be found predominantly among young objects in the Galaxy; indeed, we (Chi et al., 1992 c) make the case for this on a statistical basis. More recently, we (Chi et al., 1992 d), have examined the longitude dependence of f_E and concluded that f_E is somewhat greater in the general direction of the Galactic Anti -centre, viz, towards the local spiral arm (the solar system is on the inside edge of this arm). Such a result is as would be expected in the light of the previous discussion, assuming, that is, that protons are common in the primary beam.

Figures 2 and 3 show the f_E values from recent studies of the Sydney data (Chi et al., 1992 e to be published); these indicate that there may be further longitude - dependent anisotropies for the particles detected in this experiment (preferentially, but not wholly, heavy nuclei).

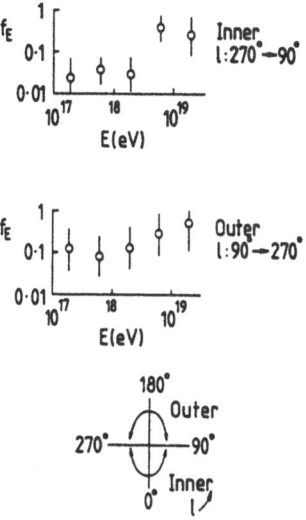

Figure 2 f_E versus energy for different longitude ranges for an analysis of the Sydney data by Chi et al. (1992 f). The longitude ranges are straighforward Inner and Outer regions.

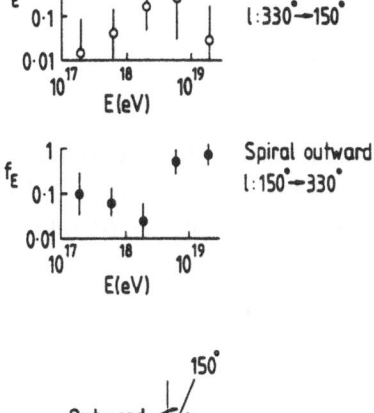

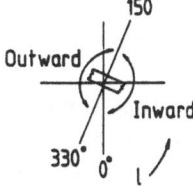

Figure 3 f_E versus energy for different longitude ranges for an analysis of the Sydney data by Chi et al. (1992 f to published). The longitude ranges relate to the direction of the local spiral arm, the reason for this choice being the presence of the Galactic magnetic field aligned along the arm. 'Inward' indicates towards the Galactic Centre along the spiral arm and 'Outward' is in the opposite direction. The magnetic field, both regular and irregular, will be of lower magnitude in the Outward direction.

2.2. The Presence of Clustering in the Arrival Directions

Detecting clustering in the arrival directions is an obvious way of searching for 'sources' but it is fraught with statistical dangers. It can be said immediately that if there were strong clustering we would have known about it for many years - and none has been reported. Nevertheless, we (Chi et al., 1992 a) have examined the world's data in an attempt to 'dig a signal out of the noise'. Our result is shown in Figure 4. The prevalence of clusters along the Galactic Plane is heartening - although their presence is not universally accepted.

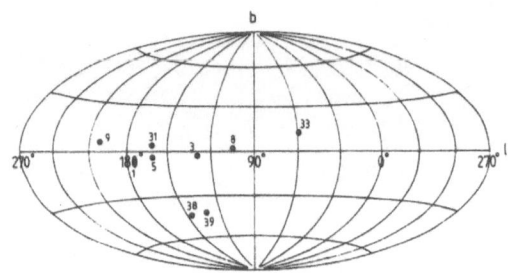

<u>Figure 4</u> Clusters of cosmic ray arrival directions for the world data above 10^{19} eV from the analysis of Chi et al. (1992 a). There is no doubt that some of the clusters are 'spurious' in the sense that they do not represent the direction of specific sources but the concentration along the Galactic Plane makes us believe that some, at least, are genuine.

An immediate question to ask is whether the actual 'sources' are in the directions of these cluster or not. The point is that the intensity enhancement in the general plane region will throw up an excess of statistical groupings. No doubt this is true but it should be pointed out that the f_E function used is considerably wider than the narrow Galactic Plane (note the exp. - b^2 factor) and the strong concentration does suggest that at least <u>some</u> of the clusters represent sources.

2.3. The Nature of the Galactic Sources

Although it is important to have shown that there are sources within the Galaxy capable of accelerating particles (protons) to energies above 10^{19} eV, even more important is the identification of such sources. We (Chi et al., 1992 c) argue that a case can be made for some at least being pulsars. The most significant cluster is towards a nearby pulsar: PSR JO157 + 6212; this pulsar is notable for having the highest surface magnetic field of any known.

A modest case can also be made for the CRAB being a source of particles of ultra-high energy, particularly when attention is directed towards particles identified as protons (by a technique to be described later) - 6 observed compared with 0.6 expected in a 6° radius 'error circle'. Figure 5 shows the energy spectrum of gamma rays from the CRAB and its extrapolation to the highest energies. It will be noted that the very tentatively claimed proton flux is of the order of that expected if (and it is a big 'if') it is permissible to extrapolate the spectrum over many orders of magnitude.

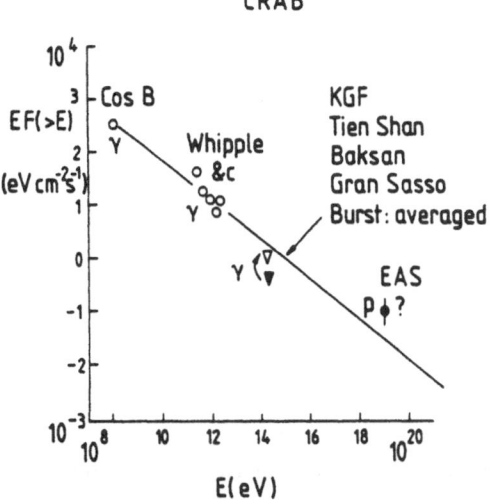

Figure 5 Energy spectrum of 'cosmic rays' from the direction of the CRAB rebula. The line drawn is an extrapolation of the spectrum from the COSB and Whipple data. The fluxes from the COSB and Whipple are well-founded. That shown as 'Burst: average' is less well so but represents the average burst flux from the detectors listed for the simultaneously detected burst on 23 February 1989 (eg. Alexeenko et al., 1991). The upper triangle represented an attempt to include other bursts which were not detected as such.

'EAS P?' represents the tentative flux that Chi et al. (1992 c) determined from their analysis of the EAS data. It is apparent that it is at least of the expected order of magnitude if the low energy spectrum is allowed to continue with unchanged exponent (it is appreciated that insofar as the line relates to γ-rays the proton flux would be expected to be higher.

2.4. THE NATURE OF THE PRIMARY PARTICLES

It has already been remarked that protons from local (d ≺ 1 kpc) sources have a chance of being deflected to such a small extent as to make their sources apparent. Such a situation is impossible for heavy nuclei. There is thus a great premium on identifying the primary masses.

It should be possible, in principle, to at least distinguish between protons and heavy nuclei by examining the lateral distribution of the EAS particles at the observation level. Figure 6 illustrates the principle.

40

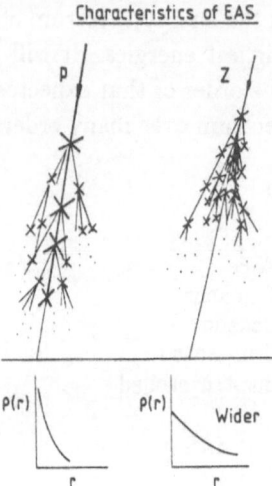

Figure 6 Schematic diagram of the lateral distributions of particles in EAS initiated by protons (P) and heavier nuclei (Z). We expect the distributions to be wider for Z for both the electron and muon components.

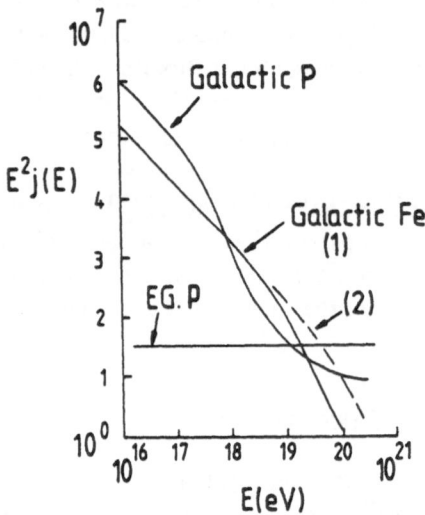

Figure 7 Energy spectra of the (assumed) two components of cosmic rays: protons and iron nuclei (after Chi et al., 1992 f). Galactic Fe (1) and (2) are variants. There can be no claim for great accuracy in this graph but it represents our preferred situation at present.

It seems that there is general agreement in principle but many would argue that the existing data are not sufficiently precise to make the distinction. We (Chi et al., 1992 **b, d, e, f, g**) have attempted to do this and Figure 7 indicates the ensuring spectra of the two components, protons (P) and heavy nuclei (Z - assumed to be iron). No doubt Figure 7 is not accurate in detail but is does at least form the basis for argument.

It will be noted that an extragalactic proton component is needed. The mechanism whereby these particles are accelerated is not known but one is free, here, to advance more exotic phenomena than are permitted in our own Galaxy. Colliding galaxies with re-connecting magnetic fields look hopeful, as do active galactic nuclei.

3. Conclusions

A reasonable case can be made for particles above 10^{19} eV coming from a variety of sources and having different masses. Certainly, it is hard to escape the conclusion that the heavy nucleus flux is significant - indeed, it is the presence of such nuclei, with their near - isotropic distribution, that causes such a problem with source identification.

Figure 8 indicated the situation schematically.

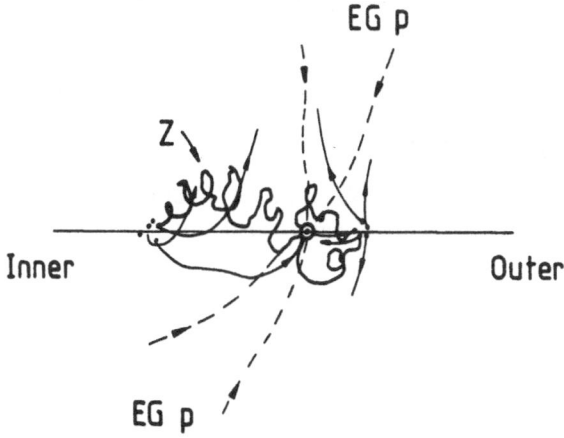

Figure 8 Pictorial representation of the trajectories of protons iron nuclei which appear to fit much of the present data. The diagram represents a side view of the Galaxy with the sun in the centre. The heavier lines represent the trajectories of heavy nuclei (Z), probably iron.

One final point that should be mentioned concerns the extragalactic 'protons'. It has been suggested recently (Bhattacharjee, 1991) that cosmic strings may be responsible for these particles, the protons being generated when the strings collapse. There is in fact a problem in that gamma rays generated at the same time as the protons should put too much energy into the electromagnetic component in the universe; however, it has recently been realised (Chi et al., 1992 **g**) that the 'protons' might conceivably be gamma rays instead. The point here is that although at 10^{15} eV p- and γ- initiated showers are very different, at 10^{19} eV or so it appears that they should be rather similar. The total electromagnetic component need not have too high a flux on this model and there are no obvious drawbacks. Further analysis of the gamma ray hypothesis is needed.

REFERENCES

Alexeenko, V.V. et al. (1991) 'Burst - like event observed by Baksan EAS array from the Crab Nebula 23 Feb. 1989', Proc. 22nd Inst. Cosmic Ray Conf., Dublin, 1, 293-296.

Bhattacharjee, P. (1990) 'Ultra-high energy particles from cosmic strings', Astrophy. Aspects of the most energetic cosmic rays, World Scientific Eds. M. Nagano and F. Takahara, 382-399.

Chi, X., Szabelski, J., Vahia, M.N., Wdowczyk, J. and Wolfendale, A.W. (1992 **a**) 'Cosmic rays of the highest energies: I. Evidence for a galactic component', J. Phys. G: Nucl. Part. Phys., 18, 539-552.

Chi, X., Vahia, M.N., Wdowczyk, J. and Wolfendale, A.W. (1992 **b**) 'Cosmic rays of the highest energies: II. The mass composition and primary spectrum', J. Phys. G: Nucl. Part. Phys., 18, 553-566.

Chi, X., Szabelski, J., Vahia, M.N., Wdowczyk, J. and Wolfendale, A.W. (1992 **c**) 'Cosmic rays of the highest energies: III. The nature of the candidate discrete sources', J. Phys. G: Nucl. Part. Phys., 18, 567-577.

Chi, X., Wdowczyk, J. and Wolfendale, A.W. (1992 **d**) 'Cosmic rays of the highest energies: IV. Further studies of the galactic component', J. Phys. G: Nucl. Part. Phys., 18, 1259-1268.

Chi, X., Wdowczyk, J. and Wolfendale, A.W. (1992 **e**) 'Cosmic rays above 10^{17} eV: I' (to be published).

Chi, X., Wdowczyk, J. and Wolfendale, A.W. (1992 **f**) 'Cosmic rays above 10^{17} eV: II anisotropies for the various mass components' (to be published).

Chi, X., Dahanayake, C. Wdowczyk, J. and Wolfendale, A.W. (1992 **g**) 'Cosmic gamma rays from collapsing cosmic strings' (to be published).

Wdowczyk, J. and Wolfendale, A.W. (1984) 'Galactic cosmic rays above 10^{18} eV', J. Phys. G: Nucl. Phys., 10, 1453-63.

HIGH ENERGY COSMIC RAYS FROM CORES OF ACTIVE GALACTIC NUCLEI

R.J. PROTHEROE and A.P. SZABO
Department of Physics and Mathematical Physics
University of Adelaide, Adelaide, South Australia 5000, Australia

ABSTRACT. We consider the possibility that shock acceleration of protons taking place in the radiation field of active galactic nuclei might result in an observable flux of high energy cosmic rays at Earth.

1. Introduction

Shock acceleration may be responsible for producing energetic particles and radiation in active galactic nuclei (AGN), possibly at a shock in an accretion flow onto a super-massive black hole (Protheroe and Kazanas 1983, Kazanas and Ellison 1986). If the acceleration takes place in the core (central region) of an AGN, where the radiation field is particularly intense, the maximum proton energy achievable is governed by proton–photon interactions (e.g., Sikora *et al.* 1987). High energy particles will be produced when accelerated protons interact via pair production and pion photoproduction with photons. In an important paper Stecker *et al.* (1991) have shown that a diffuse flux of high energy neutrinos produced in this way from unresolved AGN may be observable with proposed neutrino telescopes (see, e.g. Barwick *et al.* 1992); our calculation of the intensity of high energy neutrinos from AGN is given elsewhere (Szabo and Protheroe 1992a).

At around 10^{19} eV the observed cosmic ray spectrum flattens, possibly due to an extra-galactic component (Shapiro and Silberberg 1983) of cosmic rays accelerated at shocks in jets of active galactic nuclei (Biermann and Strittmatter 1987, Ip and Axford 1991, Rachen and Biermann 1992). At energies up to about 10^{15} eV, the bulk of the comic rays are assumed to result from first order Fermi acceleration at supernova shocks in the interstellar medium. There are difficulties in accelerating cosmic rays with supernova shocks much above this energy, and between 10^{15} eV and 10^{19} eV the origin is uncertain.

In this paper, we consider whether cosmic rays above 10^{15} eV could also be produced in AGN. We use the spectra of high energy neutrons produced during and after shock acceleration by interactions of protons in a radiation field appropriate to AGN to calculate the contribution to the extragalactic cosmic ray pool from individual AGN. Using this result, together with the X-ray luminosity function, we make predictions of the expected cosmic ray intensity from unresolved AGN.

M. M. Shapiro et al. (eds.), Particle Astrophysics and Cosmology, 43–51.
© *1993 Kluwer Academic Publishers.*

2. The AGN Model

The AGN model we adopt is that described by Protheroe and Kazanas (1983) and developed by Kazanas and Ellison (1986). A shock at radius $R = x_1 R_S$, where R_S is the Schwarzschild radius, is assumed to develop in an accretion flow onto a supermassive black hole and be supported by the pressure of relativistic particles. If the radiation from the central region is optically thick, we may make the approximation that it has the same relation between flux at the surface, F, and brightness, B, as for black body radiation, $B = F/\pi$. In this case, the radiation density in the vicinity of the shock is related to the infrared to hard X-ray AGN continuum luminosity, L_C, by

$$U_{\rm rad} \simeq L_C / \pi R^2 c, \tag{1}$$

assuming the central region coincides with the region inside the shock.

The matter density at the shock is related through the accretion rate to the luminosity. We find the number density in the accreting plasma at the shock to be

$$n_{\rm p} \simeq U_{\rm rad} c / m_{\rm p} u_1^3 Q \tag{2}$$

where Q is the efficiency of conversion of bulk kinetic energy of accreting plasma into energetic particles at the shock of, and $u_1 = x_1^{-1/2} c$ is the upstream flow velocity. Equations (10) and (17) of Kazanas and Ellison (1986) can be used to obtain $Q = (1 - 0.1 x_1^{0.31})$.

Finally, the black hole mass is proportional to luminosity, and from Figure 5 of Kazanas and Ellison (1986) we obtain

$$\frac{M}{M_\odot} \simeq 10^{-38} x_1 \left(\frac{L_C}{\rm erg\ s^{-1}} \right). \tag{3}$$

The available data (Wandel and Yahil, 1985), after applying a bolometric correction of $\times 5$ to convert approximately the luminosities in the blue region of the spectrum to L_C, suggest that x_1 values in the range 10 to 100 appear consistent with the model.

Many of the observed AGN continuum spectra display characteristics which may be roughly split into two categories: (a) a spectrum with negligible energy density in the infrared component from the central region of the AGN and an $\epsilon^{-1.7}$ photon spectrum above the UV bump, and (b) a flat quasar–like spectrum which has roughly equal energy per decade from the infrared to hard X-rays, except in the UV region. We will take these model AGN continuum spectra, with equal energy density in the power–law and black body components, to bracket the spectra in the central regions of AGN.

3. Shock Acceleration

A simple idealized picture of shock acceleration (see, e.g., Gaisser 1990) in the absence of energy losses is as follows. The upstream and downstream plasma have flow velocities u_1 and u_2 respectively where, for strong shocks, $u_1/u_2 = 4$. The motion of high energy particles in the local plasma frame may be approximated by diffusion with diffusion coefficients D_1 and D_2 in the upstream and downstream regions. A constant fractional energy gain, ξ, occurs for every acceleration cycle. Particles crossing the shock from downstream to upstream will eventually be convected back to the shock, whereas particles crossing from upstream to downstream have some probability, P_{esc}, of escaping downstream, never to return to the shock. For strong shocks, $P_{esc} = \xi$. A consideration of convective transport towards and away from an infinite plane shock gives the cycle time for particles to propagate upstream from the shock, be convected back across the shock and return to the shock,

$$T_{cycle} = \frac{4}{c} \left(\frac{D_1}{u_1} + \frac{D_2}{u_2} \right). \tag{4}$$

The acceleration rate is then simply,

$$\left. \frac{dE_p}{dt} \right|_{accn} = \frac{\xi E_p}{T_{cycle}}. \tag{5}$$

Provided the diffusion coefficients are proportional to E_p then protons are accelerated at a constant rate while in the acceleration region. The time scale for escape from the acceleration region is $T_{esc} = T_{cycle}/P_{esc}$.

Making the approximation that $D_1 \simeq D_2 = \frac{1}{3} b r_g c$ where r_g is the gyroradius and $b \geq 1$ is a constant, we find that

$$\left. \frac{dE_p}{dt} \right|_{accn} = \frac{1.35 \times 10^{12}}{b} \left(\frac{B}{\text{gauss}} \right) \beta_1^2 \;\; \text{eV s}^{-1} \tag{6}$$

where $\beta_1 = u_1/c$, and B is the magnetic field in the region of the shock. We obtain the maximum proton energy by assuming that this occurs where

$$\left. \frac{dE_p}{dt} \right|_{accn} = - \left. \frac{dE_p}{dt} \right|_{p\gamma} \tag{7}$$

where $(dE_p/dt)|_{p\gamma}$ is the energy loss rate due to proton–photon collisions.

4. The Simulation

The Monte Carlo method is used to follow a proton as it accelerates and to model its interactions (Szabo and Protheroe 1991). After acceleration, protons are assumed to be trapped in the central region and interact repeatedly with the radiation field. The cross section for pair production is compared with the photoproduction cross section in Figure 1. For detailed modeling of pair production interactions we use the differential cross sections given by Motz et al. (1969) for the case of negligible nuclear recoil. An energy and direction for the positron is sampled in the proton rest frame using the rejection technique, and then Lorentz transformed to the laboratory frame. The energy of the electron and the final energy of the proton are obtained by energy and momentum conservation.

For photoproduction interactions, we use exclusive data of Genzel et al. (1973) near threshold, and inclusive data of Moffeit et al. (1972) at higher energies. Exclusive interactions are modeled exactly in the centre of momentum frame using the rejection method. For inclusive interactions we assume Feynman scaling to be approximately valid (for details of the technique see Szabo and Protheroe 1992b). The energy loss rate of protons is shown

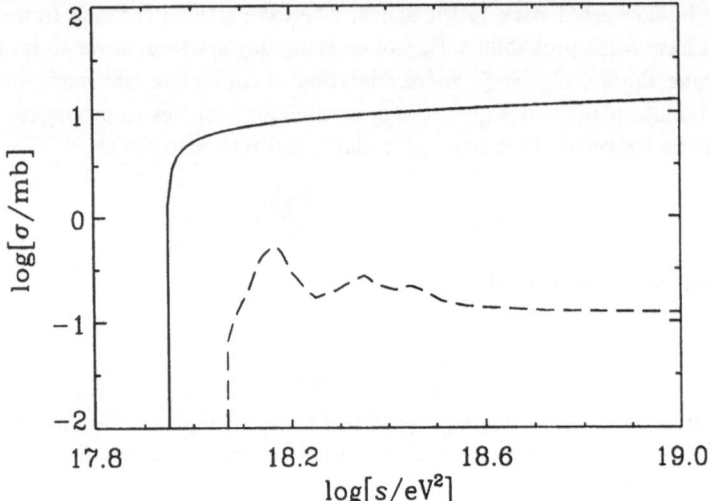

Figure 1. Total cross section for pion photoproduction (dashed line) compared with that for pair production (solid line) as a function of the center of momentum frame energy squared, s. (Pair production total cross section is from Maximon 1968.)

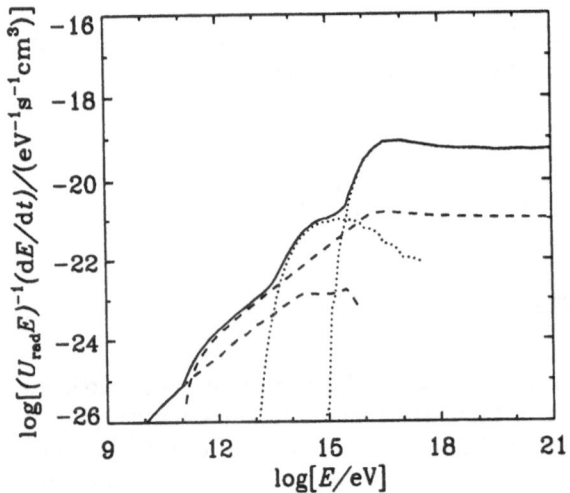

Figure 2. Energy loss rates of protons due to $p\gamma$ collisions for spectrum (a). Contributions from interactions with UV bump photons (dotted lines) and photons from the power-law part of the spectrum (dashed lines) are shown together with the total (solid lines). For each component, the lower energy curves are for pair production and the higher energy curves are for pion production.

in Figure 2 for spectrum (a). Contributions from pair production and pion production are shown separately for interactions with the power–law component (dashed lines) and black body component (dotted lines) of the field; curves for the lower energy threshold are for pair production.

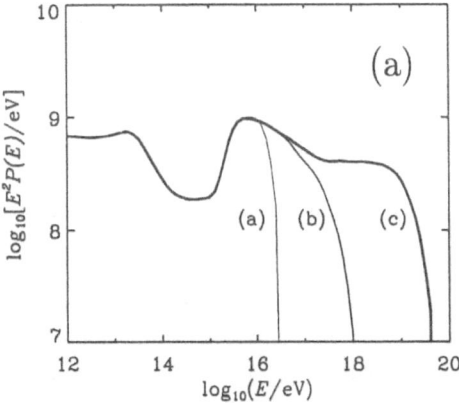 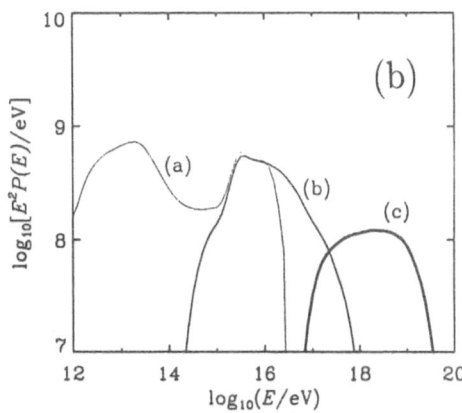

Figure 3. (a) Spectra of neutrons produced in an AGN per low energy proton injected into the accelerator. (b) Spectra of cosmic ray protons escaping from an AGN per low energy proton injected into the accelerator. In each case, $P(E)dE$ give the number in the range E to $(E+dE)$ per injected proton. Results are multiplied by E^2, and are given for X-ray luminosities $L_X = 10^{42}$ (curves a), 10^{45} (curves b), and 10^{48} erg s^{-1} (curves c), and apply to $x_1 = 30$. The reduction at high energies in part (b) is due to interactions of neutrons with photons during escape from the central region, while the reduction at low energies is due to interactions of protons from neutron decay with protons in the accreting plasma. To obtain the matter densities, we have assumed spherical accretion; non-spherical accretion could give rise to more protons escaping from the AGN at lower energies.

As a result of interactions during and after acceleration, secondary particles will be produced. Assuming accelerated protons are trapped in the central region, the secondaries will include $e^\pm$ from pair production and from $\pi - \mu - e$ decay, ν's from $\pi - \mu - e$ decay, γ-rays from π^0 decay, and neutrons. We assume that secondary $e^\pm$ and γ-rays will cascade in the radiation field, and ultimately produce the observed AGN continuum. From our simulations, the energy going into $e^\pm$ and γ-rays per injected proton is $W_{e\gamma} = 7 - 10$ GeV for $E_{\max} = 10^{15} - 10^{19}$ eV.

We show in Figure 3(a) the spectrum of neutrons produced as a result of injecting one proton into the accelerator for the case where $b = 1$ (Bohm diffusion coefficient) and $x_1 = 30$ for various luminosities. To obtain the spectrum of cosmic rays escaping per injected proton we must consider the fate of these neutrons. They are themselves subject to pion photoproduction interactions. However, not being trapped magnetically some fraction will escape from the central region. Taking account of the anisotropy in the radiation field, we have calculated the escape probability of neutrons produced near the shock. Having escaped from the intense radiation field of the central region, a relativistic neutron will decay on average after traveling a distance $r_0 \simeq 2.8 \times 10^4$ (E/eV) cm. The resulting proton will then diffuse in the magnetic field which is tied to the accreting plasma. We assume the pressure in magnetic turbulence tracks the plasma pressure, giving the diffusion coefficient at radius r

$$D(r) = D(R)(r/R)^{5/2} \tag{8}$$

where $D(R)$ is the diffusion coefficient at the shock discussed earlier. Protons will be trapped in the accreting plasma for a time $t_{esc} \sim r^2/2D(r)$. Thus the ratio of escape to interaction times is

$$t_{esc}/t_{pp} \simeq n\sigma_{pp}cr^2/2D(r) \tag{9}$$

where σ_{pp} is the pp inelastic cross section, and n is the density of nuclei in the accreting matter.

We use Equation (2) to obtain the number density of nuclei in the accreting plasma as a function of radius, luminosity, and x_1, and we use this to obtain t_{esc}/t_{pp} from which we estimate the probability of a proton from neutron decay surviving pp collisions. Having escaped from the enhanced density region of the accretion flow, we expect the protons will readily escape from the host galaxy as extragalactic cosmic rays. We show in Figure 3(b) the spectrum of cosmic rays escaping from AGN as a result of injecting one proton into the accelerator for the case where $b = 1$ (Bohm diffusion coefficient) and $x_1 = 30$ for various luminosities.

5. Cosmic Ray Output of Individual Active Galactic Nuclei

For our results to be useful, we need to be able to scale our results on escaping cosmic rays per injected proton to obtain the output of an entire AGN. The differential cosmic ray luminosity for an individual active galactic nucleus of given 2 – 10 keV X-ray luminosity, L_X, is given by

$$\frac{dL_{CR}}{dE}\{E, L_X\} = \dot{N}_p\{E_{max}(L_X)\}E\,P\{E, E_{max}(L_X)\} \tag{10}$$

where $P\{E, E_{max}(L_X)\}$ is shown in Figure 3(b) for $x = 30$ and three values of the 2 – 10 keV X–ray luminosity, L_X. We assume the infrared to X–ray continuum is due to cascading of the $e^{\pm}$ and γ–rays produced during and after acceleration. Here we relate the injection rate of protons to the 2 – 10 keV luminosity

$$\dot{N}_p\{E_{max}(L_X)\} = \frac{k_C L_X}{W_{e\gamma}\{E_{max}(L_X)\}}, \tag{11}$$

where k_C is the ratio of the total luminosity in the infrared – hard X-ray bands to the 2 – 10 keV X-ray luminosity. Hence,

$$\frac{dL_{CR}}{dE}\{E, L_X\} = 6.4 \times 10^{11} k_C \left(\frac{L_X}{erg\,s^{-1}}\right)\left(\frac{W_{e\gamma}\{E_{max}(L_X)\}}{eV}\right)^{-1} E\,P\{E, E_{max}(L_X)\}\ s^{-1}. \tag{12}$$

If cosmic rays traveled in straight lines, we could use Equation (12) together with the luminosity distance to an AGN to calculate the cosmic ray flux from that AGN. Although it makes no sense to calculate the flux from individual AGN in this way, the description in the next section of the calculation of the intensity from unresolved AGN which is based on the results in this section will apply also to cosmic rays provided the intergalactic diffusion coefficient is sufficiently high.

6. Cosmic Ray Intensity from Active Galactic Nuclei

Given the local 2 – 10 keV X–ray luminosity function of active galactic nuclei (in cm^{-3} erg^{-1} s),

$$\rho_0(L_X) = \left.\frac{dn_{AGN}}{dL_X}\right|_{z=0}, \tag{13}$$

how the luminosity function evolves with redshift z, and the differential luminosity, dL_{CR}/dE (eV s^{-1} eV^{-1}), it is straightforward to calculate the expected cosmic ray intensity. The luminosity function per unit volume of co-moving coordinate space at redshift z is

$$\rho(L_X, z) = R_0^3 \frac{g(z)}{f(z)} \rho_0\left(\frac{L_X}{f(z)}\right) \tag{14}$$

where R_0 is the present scale size of the universe, and f and g describe the evolution of luminosity and number density in co-moving coordinate space, respectively. Integrating over luminosity and redshift, we obtain

$$\frac{dI_{CR}}{dE} = \frac{1}{4\pi}\frac{c}{H_0}E^{-1}\int dL_X \int_0^{z_{max}} dz \frac{g(z)}{f(z)}\rho_0\left(\frac{L_X}{f(z)}\right)(1+z)^{-\alpha}\frac{dL_{CR}}{dE}\{(1+z)E, L_X\}, \tag{15}$$

where $\alpha = 5/2$ for the Einstein-de Sitter model ($q_0 = 0.5$) while for the Milne model ($q_0 = 0$) $\alpha = 2$. We also integrate over x_1 assuming the distribution in $\log x_1$ is constant for $x_1 = 10$ – 100.

Equation (15) does not depend on cosmic rays traveling in straight lines. This is because all of the effects of redshifting, etc., depend on the time when the particles were produced rather than the distance. The only requirement is that within the age of the universe, or rather the time over which AGN have been producing cosmic rays, cosmic rays will have had time to diffuse to us from a sufficiently large sample of AGN such that the integration over the luminosity function in Equation (15) applies. Assuming this to be the case, we have obtained the intensity of high energy cosmic rays for the assumed AGN continuum spectrum (Protheroe and Szabo 1992). For the luminosity function, we have used the pure luminosity evolution model fits obtained by Morisawa et al. (1990). We take the values of H_0, q_0 and z_{max} given in the paper appropriate to the model used. Our results on the cosmic ray spectrum are shown in Figure 4 for $b = 1$ and $b = 10$ for spectrum (a) where they are compared with a recent survey of the observations by Stanev (1992). To show the sensitivity of our result to the assumptions, we also plot results for spectrum (b). The maximum proton energies for this case are lower, and this is because of interactions with IR photons during acceleration.

Note that our predictions shown in Figure 4 have not been normalized in any way to the cosmic ray data and are determined solely by the accretion/shock acceleration model used and by the observed X-ray luminosity function of AGN. We find it quite remarkable that the predicted contribution from AGN to the cosmic ray spectrum is of the same order of magnitude as the observed intensity in the region of the knee and higher energies. Minor adjustments to the model, for example by increasing the ratio of L_C to L_X by ~ 2 in the assumed AGN continuum, could be made to give even better agreement with the observations at 10^{16} eV.

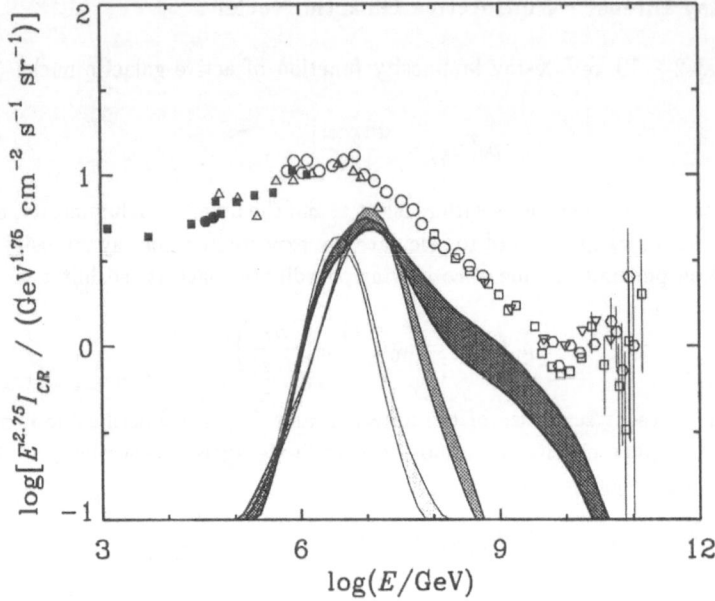

Figure 4. Predicted cosmic ray spectrum due to acceleration in AGN. The hatched bands give the range between the lowest and highest intensities obtained for the four models of the AGN luminosity function considered (Morisawa *et al.* 1990). Results are shown for the AGN continuum spectrum adopted (deficient in IR at the shock) for a ratio of scattering mean free path to gyroradius of $b = 1$ (heavy shaded band) and $b = 10$ (intermediate shading). Results for an alternative AGN continuum, with enhanced IR density at the shock, and $b = 1$ are also given (light shading). Observations are from the survey by Stanev (1992). No normalization to cosmic ray data has been made.

7. Conclusion

We have calculated the spectrum of cosmic rays produced in AGN per low energy proton injected into the accelerator. We have described how to use the observed infrared to hard X–ray continuum to normalize the predicted spectrum of cosmic rays escaping individual AGN to give the expected cosmic ray intensity from unresolved AGN.

We conclude that AGN may be an important source of cosmic rays in the region of the knee. The enhancement which appears to be present at the knee may be due to this extragalactic component. Although we have only considered the acceleration of protons in the AGN model, heavier nuclei will also be accelerated. However, no heavy nuclei will escape from the central region as they will be broken up in interactions with photons in the central region. Thus, any extragalactic component in the region of the knee will be 100% protons. If this model is correct, then one would expect to observe an enhancement in the relative abundance of protons in the cosmic rays at $\sim 10^{16}$ eV. At present the data in this energy range are indirect, being based on air shower data, and their interpretation is controversial. However, first indications from a recent study of the cosmic ray composition using the MACRO detector at Gran Sasso (Ahlen *et al.* 1992) favour a light composition at energies up to several thousand TeV. Future experiments may well be able to measure the composition more directly.

Acknowledgments

R.J.P. is grateful to T. Stanev and P.L. Biermann for discussions. A.P.S. acknowledges receipt of an Australian Postgraduate Research Award. This work is supported by a grant from the Australian Research Council.

References

Ahlen, S., *et al.*, MACRO Collaboration, Report No. LNGS–92/25, unpublished (1992).
Barwick, S., *et al.*, J. Phys. G: Nucl. Part. Phys., **18**, 225 (1992).
Biermann, P.L., and Strittmatter, P.A., Ap. J. **322**, 643 (1987).
Gaisser, T.K., *Cosmic Rays and Particle Physics*, (Cambridge University Press, 1990).
Genzel, H., Joos, P., and Pfeil, W., Landolt–Bornstein, **8**, ed. H. Schopper, (Berlin-
    -Heidelberg–New York: Springer Verlag) (1973).
Ip, W.-H., and Axford, W.I., in *Astrophysics of the Most Energetic Cosmic Rays*, edited by
    M. Nagano and F. Takahara, p. 273 (World Scientific, Singapore, 1991).
Kazanas, D., and Ellison, D.C., Ap.J., **304**, 178 (1986).
Maximon, L.C., J. Res. Natl. Bur. Std., **72B**, 79 (1968).
Moffeit, K.C., *et al.*, Phys. Rev. D, **5**, 1603 (1972).
Morisawa, K., *et al.*, Astron. Astrophys. **236**, 299 (1990).
Motz, J.W., Olsen, H.A., and Koch, H.W., Rev. Mod. Phys., **41** , 581 (1969).
Protheroe, R.J., and Kazanas, D., Ap.J., **265**, 620 (1983).
Protheroe, R.J., and Szabo, A.P., in *Proc. IVth Rencontre de Blois*, ed. J. Tran Than Van
    (Editions Frontieres, Gif sur Yvette), in press (1992).
Rachen, J., and Biermann, P.L., in *Particle Acceleration in Cosmic Plasmas*, edited by G.P.
    Zank and T.K. Gaisser (American Institute of Physics, New York) in press (1992).
Shapiro, M.M., and Silberberg, R., Ap. J. **265** 570 (1983).
Sikora, M., Begelman, M.C., and Rudak, B., Ap. J. **341**, L33 (1989).
Stanev, T., in *Particle Acceleration in Cosmic Plasmas*, edited by G.P. Zank and
    T.K. Gaisser (American Institute of Physics, New York) in press (1992).
Stecker, F.W., *et al.*, Phys. Rev. Lett., **66**, 2697 (1991a).
Szabo, A.P., and Protheroe, R.J., Proc. 22nd Int. Cosmic Ray Conf., Dublin, **2**, 380
    (1991).
Szabo, A.P., and Protheroe, R.J., in *High Energy Neutrino Astrophysics*, edited by
    V.J. Stenger *et al.* (World Scientific, Singapore), in press (1992a).
Szabo, A.P., and Protheroe, R.J., in *Particle Acceleration in Cosmic Plasmas*,
    eds. G.P. Zank and T.K. Gaisser, American Inst. of Physics, New York,
    in press (1992b).
Wandel, A., and Yahil, A., Ap. J. **295**, L1 (1985).

NEUTRINO AND GAMMA-RAY ASTROPHYSICS

R. SILBERBERG
Universities Space Research Association
Washington, DC 20024

M. M. SHAPIRO
University of Maryland
College Park, MD 20472

and C. H. STARR
Computer Sciences Corporation
Calverton, MD 20705

ABSTRACT. Neutrinos and gamma rays are complementary probes of high-energy reactions in astrophysics. The complementarity is due to the respective natures of their interactions: weak vs. electromagnetic. Solar neutrinos probe the nucleosynthesis reactions at the solar core, and possible elementary interactions like neutrino oscillations. Solar gamma rays probe the conversion of magnetic energy at the solar surface into high-energy nuclear particles and electrons during solar flares.   In supernovae, neutrinos probe the neutronization and electron pair annihilation reactions at the stellar core as the star collapses to a neutron star, while gamma rays probe the nucleosynthesis reactions along the nucleosynthesis pathway from ^{28}Si to ^{56}Ni and ^{57}Ni after these nuclei have reached the surface. In Active Galactic Nuclei the acceleration processes at the ergospheres of ultra-massive black holes and Comptonization processes have recently been observed with the Compton GRO, at energies up to 10^4 MeV. In a few years, the DUMAND array should observe neutrinos from the AGN near and above energies of 10^6 MeV, produced from protons in intense photon fields via photo-pion production and pion decay. The high energy-density environment and the acceleration processes at the AGN (e.g. at quasars, BL Lac objects, Seyfert galaxies and radiogalaxies) will be better understood by the combined exploration of gamma ray and neutrino detectors.

1. Introduction

The field of high-energy astrophysics or astroparticle physics is one of the newest and most rapidly developing branches of science. Nuclear physics, elementary particle physics and astrophysics become intertwined in this discipline.   Moreover, the objects of study in this field are largely recent discoveries, mainly in the latter half

53

M. M. Shapiro et al. (eds.), Particle Astrophysics and Cosmology, 53–94.
© 1993 *Kluwer Academic Publishers.*

of the 20th century.   Among these objects are supernovae, pulsars, accreting neutron stars in binary systems, black holes and active galactic nuclei (AGN) which are probably powered by ultra-massive black holes of about 10^8 $M_\odot$.   The most powerful of the AGN, the quasars, emit $\sim 10^{47}$ or 10^{48} ergs/sec.

Some of the physical processes occurring at such sites can be studied through their gamma-ray and neutrino emissions.   These processes include: particle acceleration, gravitational collapse, nucleosynthesis, and the effects of strong, weak, and electromagnetic interactions.

Observational gamma ray astronomy dates back only 25 years. Clark et al. (1968) discovered an enhancement of gamma rays from the direction of the Galactic center with a scintillator telescope on the satellite OSO 3.   Solar neutrinos were observed by Davis et al. (1977).   Their low abundance has been puzzling.   The favorite explanation is neutrino (ν_e) oscillations into ν_μ and or ν_τ. The detection of neutrinos from outside the solar system dates back only 5 years. At that time the proton decay detectors of the IMB and Kamiokande groups observed neutrinos from the supernova SN1987A in the Large Magellanic Cloud (Bionta et al. 1987 and Hirata et al. 1987). These neutrinos, from the collapsing stellar core, had energies near 10 MeV. Observation of high-energy neutrinos (of $\sim 10^6$ MeV) is still in the future; the Deep Underwater Muon and Neutrino Detector DUMAND, now under construction, may be observing these by 1995 or sooner. Hence, both neutrino and gamma-ray astrophysics are active and promising research fields for current graduate students.

In this article we discuss in sequential sections: (a) The complementary role of gamma rays and neutrinos for exploring the astrophysical sites and processes. This symbiosis is due to different interactions of gamma rays and neutrinos: electromagnetic vs. weak interactions. (b) Probable sites for origin of neutrinos, and methods for their observation.   (c) Observations in gamma-ray astronomy, techniques of observation, and associated problems.

2. Gamma Rays and Neutrinos as Complementary Probes in Astrophysics.

There are similarities as well as differences in production of gamma rays and neutrinos at sites of astrophysical interest that depend on the physical conditions at these sites of origin. For example, the production of gamma rays can be due to interactions of high-energy electrons in magnetic fields.   However, both gamma rays and neutrinos are produced in high-energy proton (or cosmic-ray) interactions in matter (and in dense photon fields) via pion and Kaon production and decay.

There are also characteristic similarities and differences in gamma-ray and neutrino interactions in dense or extended regions of matter or photons at or near the source regions. For example, both gamma rays and neutrinos interact (though at vastly different rates) in the collapsing supernova core, and both gamma rays and neutrinos from a neutron star can heat up a binary (giant) companion of the neutron star. In somewhat less dense regions only the gamma rays are absorbed or degraded in energy.

3. Supernovae, Pulsars, Supernova Remnants, and Accreting Neutron Stars as Sources of Gamma Rays and Neutrinos.

We shall first review briefly stellar evolution to the supernova phase and neutron star (pulsar) formation. Very massive stars with initial masses greater than about 10 $M_\odot$, i.e. > 10 solar masses, evolve into supernovae of type II. (An extremely massive star can even evolve into a black hole). Somewhat less massive stars in binary stellar systems, in which one has evolved into a dense, electron-degenerate white dwarf star that accretes mass from its binary companion can also become a supernova (type Ia). Type I supernovae have optical spectra that are nearly absent in hydrogen, due to nucleosynthesis (hydrogen burning) during collapse to the white dwarf. White dwarfs have densities of about 10^5 g/cm^3.

The stars are maintained in equilibrium against gravitational collapse by outward radiation pressure generated initially by hydrogen burning into helium with

$$p + p \rightarrow {}^2H + e^+ + \nu_e$$

as the initial step. We note that a neutrino is produced, i.e. the above reaction is a weak interaction, which permits a long hydrogen burning phase of about 10^{10} years for a star of solar mass. This has saved stars (including our sun) from rapid explosive burning, and made life on earth possible. More massive stars burn more rapidly to balance the greater gravitational inward pull, e.g. a star of 100 M_0 burns up its hydrogen fuel in 10^5 or 10^6 years. Massive stars attain higher temperatures and burn hydrogen via the catalytic CNO cycle, in which also most C and some O burn into N.

The next step, after hydrogen has burned up in the stellar core, is helium burning

$$^4He + {}^4He \rightarrow {}^8Be$$

Here ^8Be is highly unstable and decays back into helium except when occasionally it merges with ^4He and becomes ^{12}C. Some ^{12}C nuclei in turn interact with ^4He, as

$$^4He + {}^{12}C \rightarrow {}^{16}O$$

As the helium core is burned up, fusion of ^{12}C starts, e.g. $^{12}C + {}^{12}C \rightarrow {}^{20}Ne + {}^4He$ (also ^{23}Mg and ^{23}Na are formed in carbon fusion). Oxygen fusion also proceeds via $^{16}O + {}^{16}O \rightarrow {}^{28}Si + {}^4He$, and ^{31}P and ^{31}S are also formed. Burning of $^{20}Ne + {}^4He$ yields ^{24}Mg.

Thus an "onion-like" series of nucleosynthesis shells are formed in a pre-supernova star, with Si and Mg burning at the core, and (O, Ne, C), He and H burning shells outside the core. Burning of $^{28}Si + {}^4He$ yields ^{32}S, this in turn ^{36}Ar, then ^{40}Ca, etc. until the stablest nucleus ^{56}Fe is formed. (The explosive nucleosynthesis process that generates ^{56}Ni on a rapid time scale is discussed a few paragraphs later). Then nuclear burning can no longer generate radiation pressure against the gravitational force, and the stellar core collapses to a neutron star, about 1.4 $M_\odot$, but only 10 km in radius, with a density of about 10^{15} g/cm^3, i.e. about 10^{38} nucleons/cm^3.

Nuclei disintegrate in the stellar core, and the reaction

$$p \rightarrow n + e^+ + \nu_e$$

is initiated. An even greater number ($\sim$ 10 x more) of neutrinos are formed in the dense $e^+e^-\gamma$ plasma in the stellar core at energies > 1 MeV, i.e. > 2 electron rest masses m. The $e^+ + e^- \rightarrow 2\gamma$ reactions occur at such a prodigious rate that the core is cooled by the occasional $e^+ + e^- \rightarrow \nu_e + \nu_e$ or $2\gamma \rightarrow \nu_e + \nu_e$ weak interaction.

The outward radiation pressure by the 10^{58} neutrinos and the explosively rapid nucleo-synthesis outside the core (occurring in a few seconds) causes the outer part of the star to explode as a supernova.

This scenario has been confirmed in the observation of neutrinos and nucleosynthetic gamma-ray lines from SN 1987A in the satellite-galaxy Large Magellanic Cloud at about 50 x 10^3 pc (or 50 Kpc). A parsec or pc is 3.26 light years; a star at 1 parsec exhibits a parallax of 1 second of arc (3600^{-1} degrees of arc) as the earth moves from one end of its orbit to the other around the sun. The star at 1 parsec then appears to move by 1 second of arc relative to the fixed stars. The neutrinos were observed with deep underground detectors (designed to record proton decay). Six neutrinos in the 20-40 MeV energy range were observed with the IMB (Irvine, Michigan, Brookhaven) detector of Bionta et al. (1987) and eleven neutrinos in the 7-36 MeV energy range (a lower detection threshold) were observed with the Kamiokande II detector of Hirata et al. (1987).

Outside the collapsing core, explosive nucleosynthesis occurs on a rapid time scale, e.g. ^{28}Si burns, adding ^{4}He nuclei until ^{56}Ni is formed. The gamma-ray lines of ^{56}Co decay into ^{56}Fe (originally produced as ^{56}Ni) at 0.847 and 1.238 MeV were observed by Matz et al. (1988) and confirmed by several other groups. About 0.075 $M_\odot$, i.e., about 3 x 10^4 earth masses of radioactive ^{56}Ni were thus formed. Because of the large number of ^{56}Ni nuclei formed, the Compton Gamma Ray Observatory could see type I supernovae out to a distance of 10^7 pc (or 10 Mpc). Type Ia supernovae can be seen to a much greater distance than type II, since the massive "blanket" around type II obscures ^{56}Co until most of it has decayed.

There are several other gamma-rays lines from radioactive nuclei, e.g. ^{57}Co and ^{44}Ti. ^{57}Co is marginally observable from a distance like that of the large Magellanic Cloud with GRO. ^{44}Ti is less abundant, hence it is observable only from our Galaxy. However, because of its long half-life of 50 years, ^{44}Ti renders Galactic supernovae observable for a couple of centuries. Though none has been observed for 3 centuries, there may be some in obscured Galactic regions which could be detected via the ^{44}Ti line (^{44}Ti decays into ^{44}Sc, which in turn decays into ^{44}Ca) with the Compton GRO.

As the stellar core collapses to a neutron star, the frequency of rotation speeds up by many orders of magnitude owing to angular momentum conservation. Periods of pulsars are thus in the range of 0.1 to 10 seconds (while millisecond pulsars have been observed, the latter have probably speeded up due to mass transfer from a binary companion). Also, the magnetic field is enhanced by many orders of magnitude, to values near 10^{12} gauss.

The high-energy neutrinos ($E > 10^{11}$ eV) from astrophysical sites (like pulsars, supernova remnants and accreting neutron stars) have not yet been recorded. Their fluxes still have to be inferred from gamma-ray observations and various plausible assumptions about the origin of neutrinos and gamma rays and modifications of gamma-ray spectra in dense fields of photons and electrons.

Until actual observations of neutrinos and their energy spectra, alternative assumptions of the origin of gamma rays and neutrinos have to be considered.

(1) Both protons and electrons are accelerated at pulsars, and the other sites listed in the previous paragraph.

(2) Mainly protons are accelerated, and electrons are secondary products of proton interactions via μ^+ decay, and $\pi^0 \to 2\gamma \to$ e-pairs.

(3) Mainly electrons are accelerated and hence the production of neutrinos is negligible.

In strong magnetic fields and regions with a high photon density, high energy electrons (also positrons) undergo large energy losses, that yield photons including gamma rays. The origin of these in the compact sources is attributed to the synchrotron-self-Compton process, which we shall explore briefly below.

Assume a pulsar power output of 10^{37} erg/s, a period of 0.05 sec, and a magnetic field of 10^{12} gauss at the surface of the neutron star of radius 10 km. The radius of the light circle (or light cylinder) then is about 2.5×10^8 cm (about 0.4 earth radii). The magnetic field at the light circle α (r^{-3}) is about 10^5 gauss. Consider synchrotron emission by electrons at the light circle. The energy of the electron is halved in time $t_{1/2}$

$$t_{1/2} = 5 \times 10^8 \left[\frac{M_e c^2}{E_e}\right]\left[\frac{1}{H_1(G)}\right]^2 \text{ sec}$$

where $H_1 = H \sin \theta$, θ = angle of electron with respect to the magnetic field Ginzburg and Syrovatskii (1964). At the distance of the light circle, an electron of energy 1 GeV will lose half its energy in about 10^{-4} sec, and at an energy of 1 TeV, in about 10^{-7} sec. The energy of electrons thus is rapidly transformed into photons. The maximum synchrotron emission occurs at an energy

$$E_{\gamma,m}(\text{eV}) = 5 \times 10^{-9} H_1 (E_e/m_e c^2)^2$$

i.e. under the above condition for H_1, at about 1.4×10^3 eV from 1 GeV electrons, and about 1.4×10^9 eV from 1 TeV electrons.

In very strong magnetic fields, as considered here, electrons are constrained to move nearly parallel to the magnetic field lines, which in turn are curved. As electrons move along these curved lines they emit curvature radiation with energy

$$E_c(eV) = \frac{2.96 \times 10^{-5}\gamma^3}{\rho_c(cm)}$$

where ρ_c is the radius of curvature of the magnetic field line and $\gamma = E_e/m_e c^2$ (Manchester and Taylor, 1977 and Ramana Murthy and Wolfendale 1986). For 1 TeV electrons and $\rho_c = 2.5 \times 10^8$, $E_c = 10^6$ eV; for 10 TeV electrons, $E_c = 10^9$ eV.

Next we consider the energy boosting of the synchrotron photons by the inverse Compton process in which photons collide with highly relativistic electrons and gain energy. The energy of the Compton boosted photon is

$$E_\gamma = \epsilon\gamma^2 \quad \text{when} \quad \epsilon\gamma \ll m_e c^2$$

and

$$E_\gamma \sim E_e \quad \text{when} \quad \epsilon\gamma \gg m_e c^2$$

where ϵ is the energy of the ambient photon. Thus, electrons with energies 10^3 or 10^6 MeV can boost photons to similar energies.

Next, consider the flux of photons across the light circle. Assume 10^{36} ergs/sec is emitted in photons 10^3 to 10^4 eV in energy, i.e. about 10^{45} per sec. Assume isotropic emission of photons across a sphere of the radius of the light circle, about 2.5×10^8 cm. The photon flux at 10^3 to 10^4 eV then is about $10^{27}/cm^2$ sec at the light circle and the photon density is about $3 \times 10^{16}/cm^3$. With such a high photon density, degradation of high-energy photons becomes probable by photon-photon interactions: $\gamma + \gamma \rightarrow e^+ + e^-$. Defining ϵ_1 and ϵ_2 as the energies of the two colliding photons, $w = (\epsilon_1, \epsilon_2)^{1/2}$, and $r_e = e^2/mc^2 = $ (the classical electron radius), the cross section for head-on collision is:

$$\sigma = \pi r_e^2 (m/w)^2 [2 \ln (2w/m) - 1]$$

for $w \gg m$, the electron mass, and

$$\sigma = \pi r_e^2 (1 - m^2/w^2)^{1/2}$$

for $w \sim m$. As an example, for collision of 10^3 eV and 10^9 eV photons

$$\sigma \sim 0.9 \pi r_e^2 \sim 2 \times 10^{-25} cm^2 \text{ or } 200 \text{ mb}.$$

With a photon density (near 10^3 eV for a decade in energy) at the light circle of a pulsar of about 10^{17} cm^{-3}, degradation of high-energy gamma rays by pair production becomes likely.

Protons also interact with photons, as

$$p + \gamma \rightarrow \Delta \rightarrow p + \pi^0 \text{ or } n + \pi^+.$$

The latter yields neutrinos via

$$\pi^+ \rightarrow \mu^+ + \nu_\mu$$

This reaction peaks at $\epsilon E_p = 0.35$ (Stecker et al. 1991), where ϵ is the photon energy in MeV and E_p the proton energy in TeV.

In the case of an accreting binary, protons can collide with nuclei in the accretion disk or in the binary companion, producing pions yielding both neutrinos and gamma rays. For an absorbing material thickness > 100 g/cm^2, the gamma rays will be greatly attenuated, while high energy (1 TeV) neutrinos are attenuated in matter about 10^{-9} times as much, and the attenuation at lower energies is even less: the cross section $\sigma \alpha E_\nu$ for $10^3 < E_\nu < 10^6$ MeV. Thus, if a beam of protons from the pulsar sweeps across the binary companion, there are two short pulses of gamma rays, with an eclipse in between. The eclipse of the neutrino flux is shorter in duration, or absent, depending on neutrino energy and stellar density.

Gamma rays up to energies of 10^{12} eV are produced in supernova remnants like the Crab nebula. Electrons of energies at least up to 10^{13} eV must thus also be produced. Acceleration of particles to such energies by electromagnetic waves generated by a pulsar (with a magnetic dipole moment that makes an angle with the rotation axis) has been proposed by Gunn and Ostriker (1969). A different model by Cheng, Ruderman and Sutherland (1976) is based on particle acceleration in the outer gaps of the pulsar magnetospheres.

Fig. 1 shows the energy distribution of gamma rays per decade of energy for the Crab pulsar (open circles) and Crab nebula (solid circles). Fig. 1 is from White and Silberberg (1991) of the previous school at Erice, who also give the references to the data.

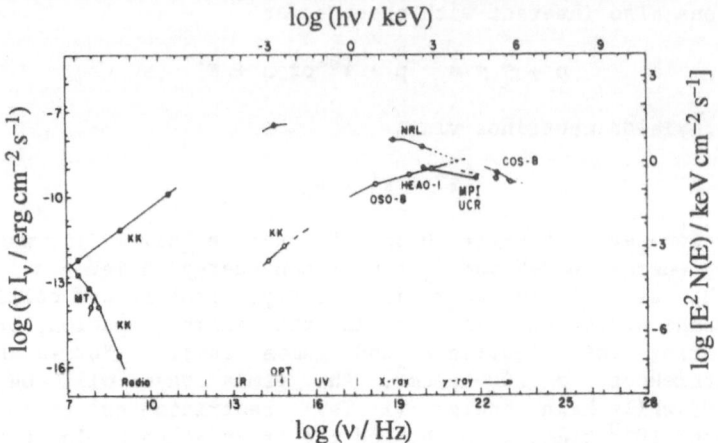

Fig. 1: Energy distribution per decade of energy for the Crab. Solid circles represent observations of the Crab nebula, open circles the Crab pulsar.

Berezinsky and Prilutsky (1976) and Berezinsky (1976) have suggested that neutrinos are copiously produced for several months by nuclear interactions of ultra-high energy protons accelerated at pulsars. These protons interact in the dense shell of the young supernova remnant and produce pions, which upon decay, yield neutrinos. Berezinsky suggested that the energy input into protons occurs at the very high rate of 10^{43} erg/s. Shapiro and Silberberg (1979) suggest that a value 10^{38} to 10^{39} erg/s is much more likely. Marscher and Brown (1978) obtained upper limits for the synchrotron (radio) power of extra-galactic supernovae. They found only one case with a measurable power output, about 2×10^{39} erg/s approximately 1 year after the supernova outburst.

4. Active Galactic Nuclei as Sources of Gamma Rays and Neutrinos.

There are several types of active galaxies. We shall briefly explore these types. They are the most powerful sites of high energy particle production, with intense photon production from radio to gamma-ray wave lengths. The power of these sources is 10^{44} to 10^{49} erg/s, i.e. up to about 10^{16} times the solar value. The source of this power is thought to be gravitational accretion upon a black hole of about 10^{8} solar masses. Most of the power is produced in the surrounding ergosphere of about 1000 light seconds, or 3×10^{13} cm. The evidence for the compact size comes from fluctuations of the X-ray flux down to time scales of 100 sec (Fabian, 1992).

Blandford and Rees (1992) estimate that the energy density in this environment is about 10^{6} erg/cm^3 and that the magnetic field strength is about 10^{4} gauss. In such an environment, electrons are rapidly degraded in energy (unless compensated by an even more rapid acceleration process). For example, a 10^9 eV electron in a 10^3 gauss field loses half its energy in 1/4 second. Also, photon-photon interactions can degrade the photon spectrum. Thus, gamma rays alone do not provide a satisfactory probe to explore the energy spectrum and intensity of particles accelerated in the black hole ergosphere.

Many processes have been proposed that yield gamma rays near ultra-massive black holes. Leiter and Kafatos (1978) have proposed Penrose processes in which a photon gains energy from a particle falling into the black hole, e.g. the Penrose Compton process in which a low-energy gamma ray scatters off an infalling electron and escapes with an energy of a few MeV. In the Penrose pair production process, a gamma ray scatters off on an infalling proton, producing pairs that can have energies of a few GeV. Acceleration processes that yield protons up to 10^{19} eV have also been proposed, e.g. by Lovelace (1976), in whose model a large potential drop is generated across the accretion disk. Another mechanism for generating energetic particles up to very high energies is acceleration by turbulent hydromagnetic shock waves in the accretion disk. This model would generate an energy spectrum of power-law form, e.g. $dJ/dE \, \alpha \, E^{-a}$ where a > 2. Stecker et al. (1991) have suggested that protons accelerated by the latter process yield high energy neutrinos, $10^{12} < E_{\nu} < 10^{18}$ eV, via the photoproduction process

$$p + \gamma \rightarrow \Delta^+ \rightarrow n + \pi^+$$

in the dense photon field, followed by pion decay into muon and neutrino. While gamma rays are subject to degradation in energy via the $\gamma + \gamma \rightarrow e^+ + e^-$ process, gamma rays up to 10^{10} eV have been observed at the blazar type quasar 3C 279 by Hartmann et al. (1992).

The active galactic nuclei fall into various classes, whose differences are due to orientation relative to the observer and to galactic type (the latter may be related to angular momentum): (1) The BL Lac objects are highly variable and without absorption lines. The jet is oriented toward the observer. (2) Blazars are a subgroup of quasars with a jet nearly toward the observer; 3C 279 and 3C 273 are examples. They can exhibit superluminous motion. (3) Quasars are extremely powerful and frequently have large red-shifts of spectral lines; values of Z = 2 and 3 are common. (4) Radio galaxies generally are elliptical galaxies with little gas or dust. Radio jets extend far out, tens of Kpc, with a shocked turbulent region at the far end. (5) Seyfert galaxies are spiral galaxies. They frequently have large turbulent clouds. There are two principal types. Seyfert 2 have the "central engine" obscured by a disk of clouds between the source and the observer. They are relatively weak sources of X-rays. A low X-ray flux is seen from NGC 1068, the closest strong Seyfert 2; these X-rays are likely to be scattered in our direction by electrons. X-rays from Seyfert 1, e.g. NGC 4151, are not obscured by the disk. They are strong sources of X-rays and low-energy (<1 MeV) gamma rays.

Much of the power output of the AGN is concentrated in the ultraviolet (the "UV bump") which is associated with radiation from the accretion disk, and in soft gamma rays near energies of 1 MeV which are associated more directly with the "central engine." The power output per logarithmic energy interval (or the luminosity spectra as a function of energy) is illustrated in Fig. 2 for three active galactic nuclei, the Galactic center, and the Galactic black hole candidate Cyg X-1. This figure is from the Proceedings of the previous Erice School, [White and Silberberg (1991)], who also give the references to the data.

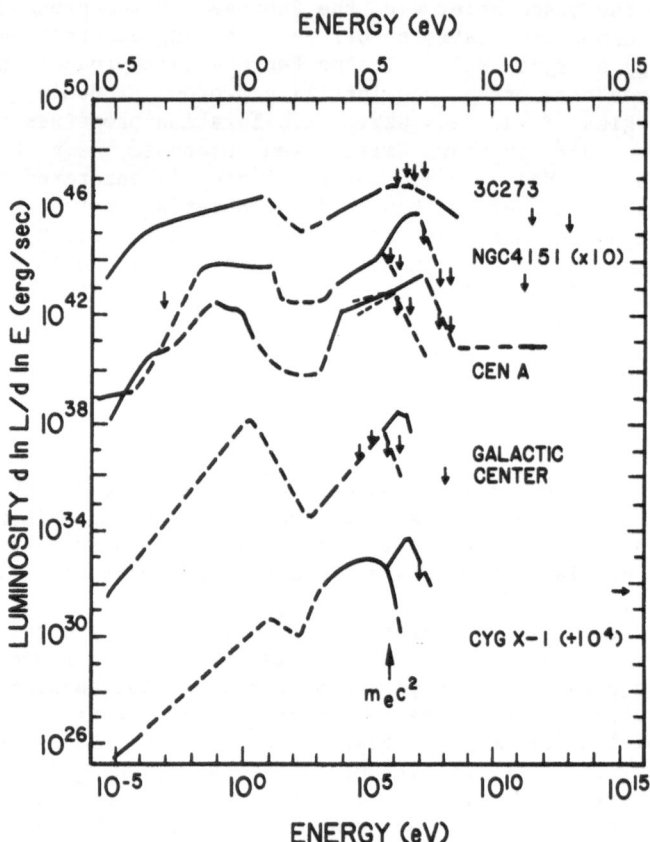

Fig. 2: Luminosities as a function of energy for active galactic nuclei. The luminosities of the Galactic center and the black hole candidate Cyg X-1 are also shown.

Recent observations from the EGRET experiment, e.g. Hartman et al. (1992) and several papers presented by the group of Fichtel et al. (1992) in the Bulletin American Physical Society, show that at certain times the luminosity of quasars of the blazar type peaks at 10^2 to 10^4 MeV, i.e. in high energy gamma rays. This is probably associated with emission of a jet from the center of the AGN. Fig. 3 from Maraschi et al. (1992) shows the luminosity spectrum of 3C 279. The lines hereare based on the synchrotron inverse Compton Model with an inhomogeneous jet. A total of 11 AGN have been observed with EGRET from 10^2 to 10^4 MeV: mostly quasars and three BL Lac objects.

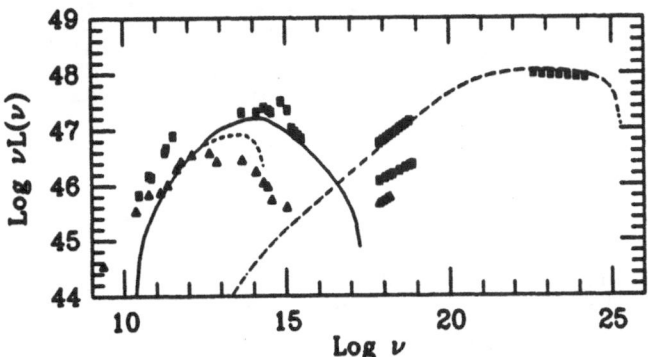

Fig. 3: The spectrum of 3C 279. Data from Makino et al. (1989), and references therein, and from Hartman et al. (1992). The inhomogeneous jet model of Maraschi et al. (1992). Fig. is from Maraschi et al.

64

Fig. 4 shows a model of the AGN as a function of R/R_G, where $R_G = 2GM/c^2$, the gravitational radius. Fig. 4 is from Collin-Souffrin (1992). It illustrates the positions of the black hole, the relativistic jets, the hot thick disc with X-ray emission, the accretion disk with the UV bump, the broad line region, the regions of high and low-ionization lines, sites of molecules and grains, and the gravitationally unstable molecular torus.

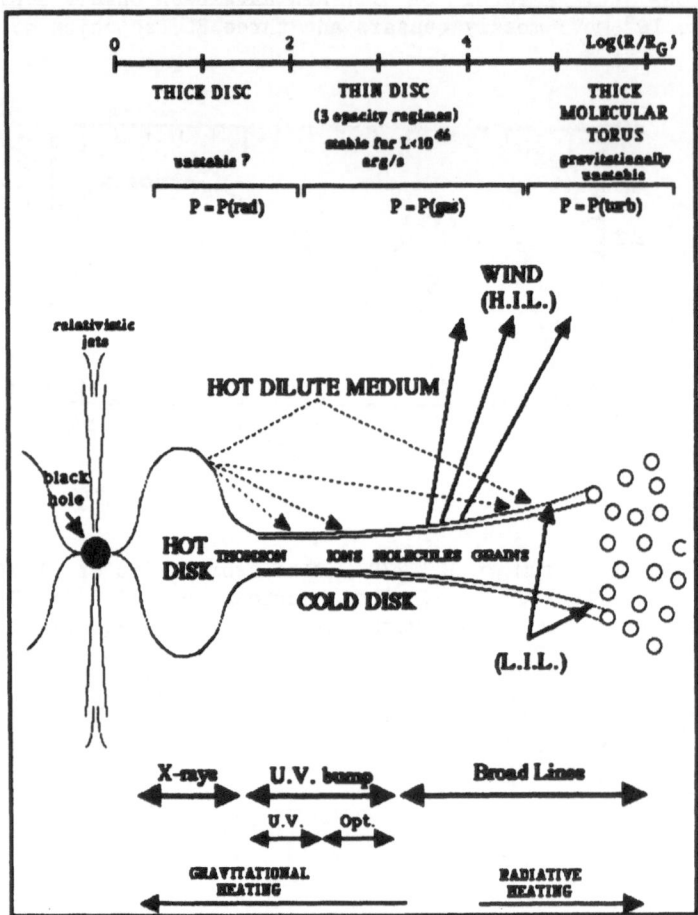

Fig. 4: The model of Collin-Souffrin for the central region of an AGN, with black hole, jets and accretion disk, with regions of X-ray, UV bump and broad line emission shown. From Collin-Souffrin (1992).

5. High Energy Neutrino Astronomy

A neutrino is detected after its conversion into a muon, which at high
energy is also associated with pion production, x pions in the example
below

$$\nu_\mu + p \rightarrow \mu^+ + n + x\pi.$$

At a neutrino energy of 10^{12} eV, the muon retains much of the energy
of the neutrino, i.e. it will be able to traverse about a kilometer of
material like water, and is collimated within about 1° of the
direction of the neutrino. This excellent collimation permits the
identification of the sites of origin.

There is a difficulty, which can be overcome: since neutrinos
are weakly interacting, only a small fraction will convert into a
charged lepton in a given detector. This problem is overcome by using
a large conversion volume for the neutrinos and a large detector
volume. This is accomplished by using a large volume of ocean water,
both for conversion of the neutrino and for transmission of the
Cerenkov light produced by the muon, in the DUMAND (Deep Underwater
Muon and Neutrino Detector) presently under construction by the
University of Hawaii and other institutes. Fig. 5 shows a drawing of
the proposed detector, 200 m high and 100 m in diameter. The small
circles show the 216 photo-multiplier tubes that record the Cerenkov
light produced by the muon. The detector has to be at a depth of
several km of ocean, where solar light does not penetrate and where
the background of muons produced by cosmic-ray interactions in the
atmosphere (via pion and Kaon decay) is greatly reduced. The
atmospheric muon background can be eliminated by looking for upward
coming (i.e. through the earth) neutrino-produced muons. Fig. 5 shows
such an upward coming neutrino-produced muon. The data are
transmitted to an on-shore laboratory for computer processing.

66

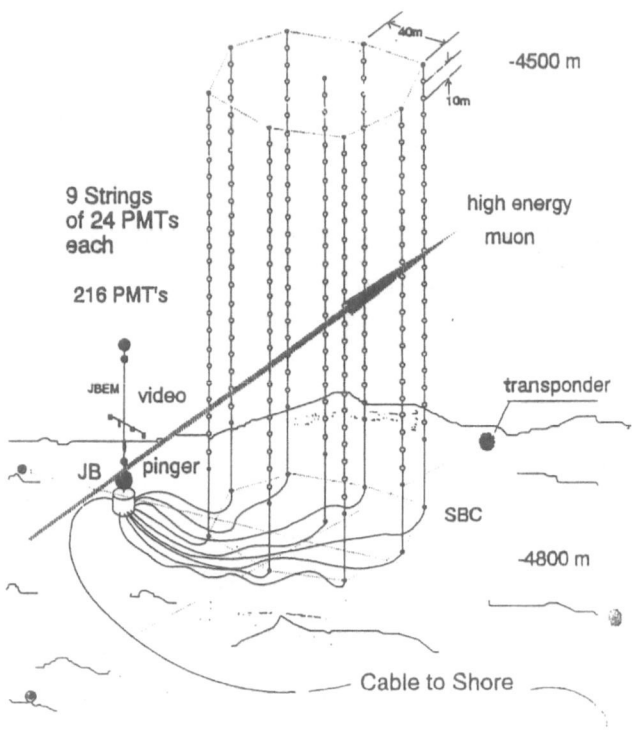

Dumand II. Hexagonal neutrino telescope.

Array dimensions: 200 m high, 100m diameter

-4500 m

9 Strings
of 24 PMTs
each

216 PMT's

high energy
muon

JBEM   video

JB   pinger

transponder

SBC

-4800 m

Cable to Shore

Fig. 5:   The DUMAND II neutrino telescope in the ocean.   An upward
(through the earth) neutrino-generated muon is shown.

Fig. 6 illustrates the conversion of a ν_μ into a muon and hadronic (mainly pion) cascade, and representative angles in units of mr. The hadronic cascade is absorbed in a few 10's of meters of water while the muon traverses about a kilometer.

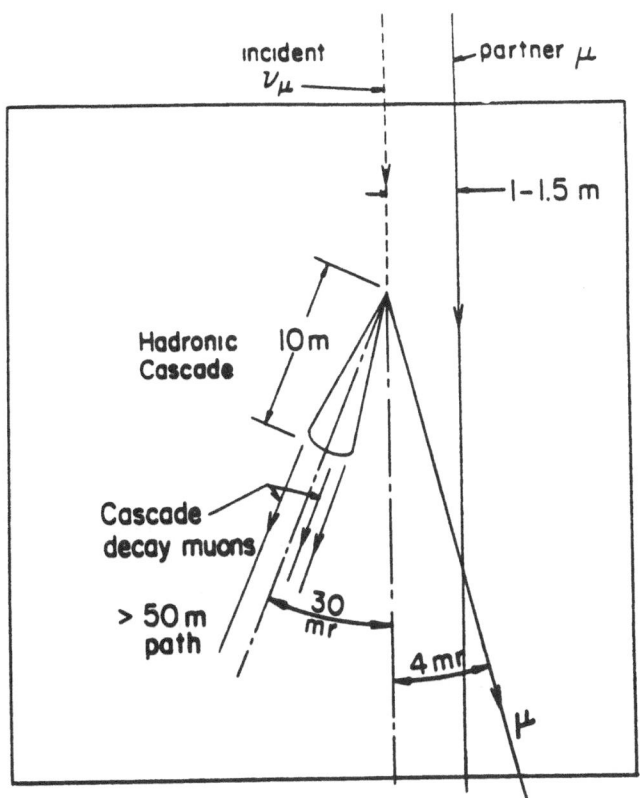

Fig. 6: Illustration of a muon neutrino interaction (at 10^{13} eV) with muon and hadronic cascade production. When produced in atmosphere it may be accompanied by a partner muon.

In Table 1 we summarize possible classes and examples of neutrino sources. There are likely to be also unanticipated sources. The classes shown are the Active Galactic Nuclei, accreting Galactic neutron stars and black holes, young supernova remnants and pulsars. In the latter, the photon fields are very intense and $p\gamma \rightarrow n \, \pi^+$ reactions, with $\pi^+ \rightarrow \mu^+ + \nu_\mu$, is likely to occur.

Table 1. Possible classes of neutrino sources

Class	Examples	Publications
Quasar	3C 273, 3C 279	Protheroe and Kazanas (1983), Kazanas and Ellison (1986), Stecker et al. (1991)
Seyfert 1	NGC 4151	Stecker et al. (1991)
Seyfert 2	NGC 1068	Silberberg and Shapiro (1979)
Radiogalaxy	Cen A, Virgo	Biermann and Strittmatter (1987)
Galactic Center		
Accreting neutron star	Cyg X-3, Her X-1, SS433	Berezinsky et al. (1986), Eichler (1981)
Accreting black hole	Cyg X-1	
Young supernova remnant		Berezinsky and Prilutsky (1976), Shapiro and Silberberg (1979)
Pulsar	Crab, Vela	This paper

We shall next explore possible neutrino fluxes. Silberberg and Shapiro (1979) explored the Seyfert 2 galaxy NGC 1068, considering it to be just like the Seyfert 1, NGC 4151, except that the central power source is obscured (due to orientation toward the earth) by intervening material. An event rate of about 10^3 per year for $E_\nu > 4$ x 10^{12} eV in an effective volume of 10^{10} m^3, i.e. a cube 2 km on side, or average area of 6 x 10^6 m^2, was estimated. The present planned detector has an area of 2 x 10^4 m^2, eventually to be expanded to 10^5 m^2. In the latter volume, our renormalized event rate is about 15/year. The recent calculations of Stanev (1992) for $E_\mu > 1$ TeV (note: $E_\mu \sim 1/2$ E_ν) yield 12 to 25, i.e. a similar value. (The difference in number of events above 2 and 4 TeV is largely balanced by $\sigma_\nu \sim E$).

The integrated value over the universe results in a diffuse neutrino flux. Berezinsky (1976) calculated this flux for the early Bright Phase of galaxy formation.   Shapiro and Silberberg (1979b) provided a somewhat more conservative estimate for active galactic nuclei. More recently, Stecker et al. (1991) evaluated the diffuse flux from the X-ray luminosity function for AGN of Morisawa and Takahara (1989), based on data from the GINGA satellite. Szabo and Protheroe (1992) found an error in the equation of Morisawa and Takahara, as a result of which the diffuse neutrino flux of Stecker et al. (1991) above 10^{15} eV has to be reduced by a factor of 20. However, near 10^{12} eV, Szabo and Protheroe (1992) obtain a higher flux; they consider it likely that the lower energy protons are confined longer, and will interact with X-ray photons (that are less numerous than the UV). They and Stanev (1992) estimate that the neutrino generated muon event rate per year in 10^5 m^2 is between 1350 and 5000, compared to an atmospheric background of 900.

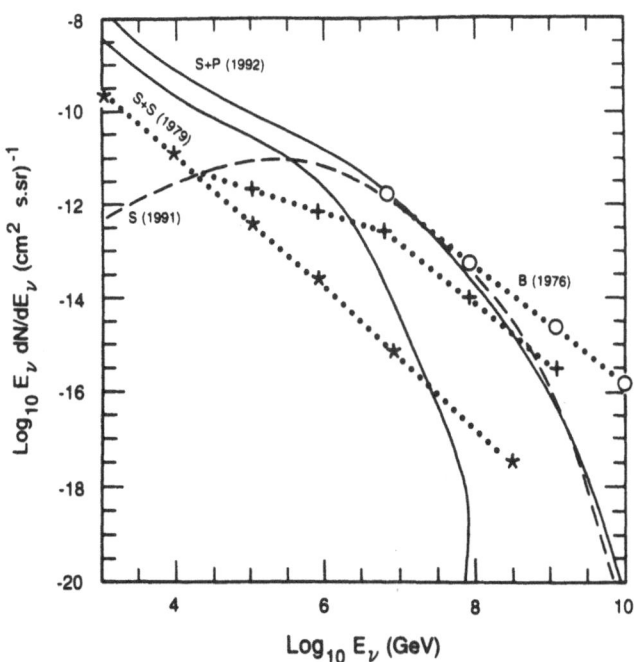

Fig. 7: The extra-galactic diffuse neutrino background as estimated by Berezinsky (1976) in the bright phase of galaxy formation shown by the line with circles; by Shapiro and Silberberg (1979), the range of values is given by the crosses and stars; Stecker (1991), as corrected by Szabo and Protheroe (1992), shown by the dashed line; and Szabo and Protheroe (1992) calculated with various assumptions, shown by the solid lines. The figure is based on Stanev (1992), with the addition of values of Berezinsky and Shapiro and Silberberg.

Fig. 7 shows a comparison of the calculations of Berezinsky (1976), Shapiro and Silberberg (1979), Stecker (1991), as corrected by Szabo and Protheroe (1992), and Stanev (1992). We note that the old calculation of Shapiro and Silberberg (1979) agrees rather well with those of Szabo and Protheroe (1992) and Stanev (1992).

6. What can be learned from Neutrino Astronomy

To summarize, neutrino astronomy allows us to learn the following:

1. The production rates of high-energy protons in AGN and compact Galactic sources, e.g. in pulsars, accreting neutrons stars and black holes.
2. The ratios of energy input into protons and electrons (when combined with gamma ray data) in AGN and compact Galactic sources.
3. The energy spectra of protons in AGN and compact Galactic sources.
4. The acceleration processes in AGN and compact Galactic sources.

7. Observational Gamma Ray Astronomy

In this and the following sections, issues relating to how observational gamma ray astronomy is performed and what observational gamma ray astronomy is currently investigating is briefly outlined. As has already been mentioned, gamma rays are produced in a variety of sites and these include solar flares, compact objects, explosive objects, nucleosynthesis events, and extragalactic sources.

Whether from balloon-borne or spacecraft platforms, observational gamma ray astronomy has primarily used scintillation detectors and germanium solid-state spectrometers to detect low-energy gamma rays, and spark chambers to detect high-energy gamma rays greater than 50 MeV. A summary of the historical development and significance of gamma ray astrophysics can be found in Kurfess 1992 and Fichtel 1991, and references therein. The material here and in the sections below comes primarily from Kurfess 1992 and from the initial papers of the Compton Gamma Ray Observatory, a recently launched NASA satellite dedicated to performing observational gamma ray astronomy over six decades of energy in the gamma-ray band from 15 keV to 20 GeV.

As described by Kurfess 1992, one of the first satellite missions devoted to observational gamma ray astronomy was the second Small Astronomy Satellite (SAS-2) which was launched by the United States in 1972 and performed a 7-month survey of the galactic plane at energies above 50 MeV. This mission was followed by COS-B, launched by the European Space Agency in 1975, which operated for seven years. SAS-2 and COS-B provided maps of the gamma-ray emission from the galactic plane and verified that this emission was primarily caused by cosmic ray interactions with the interstellar medium, with the emission above 100 MeV being dominated by gamma rays produced by the decay of neutral pions generated by the interaction of protons with the interstellar medium, and with the emission below 100 MeV largely due to bremsstrahlung resulting from the interaction of cosmic ray electrons with the interstellar medium and by inverse Compton radiation. Two other gamma ray experiments were later launched as part of the NASA High Energy Astronomical Observatory (HEAO) series: HEAO-1 and HEAO-3.

HEAO-1 (Matteson et al. 1978) carried a scintillation spectrometer and HEAO-3 (Mahoney et al. 1980) carried a germanium spectrometer, which allowed high resolution spectroscopy. Both performed a full sky survey in the 10 keV - 10 MeV energy range. Numerous balloon-borne experiments have also been flown to observe the gamma-ray sky. While long-duration balloon experiments have flown, the duration of balloon observations is typically limited to several hours (or a couple of days at best).

8. The Compton Gamma Ray Observatory (GRO)

One of the most recent spacecraft platforms for observational gamma ray astronomy is the Compton Gamma Ray Observatory (GRO), which was launched into a 400-km earth orbit by NASA on 5 April 1991. The Compton Observatory consists of four instruments that together cover an energy range from 15 keV to 20 GeV. The objectives of the Compton Observatory include a full sky survey above 1 MeV at a sensitivity 10-20 times greater than previous experiments.

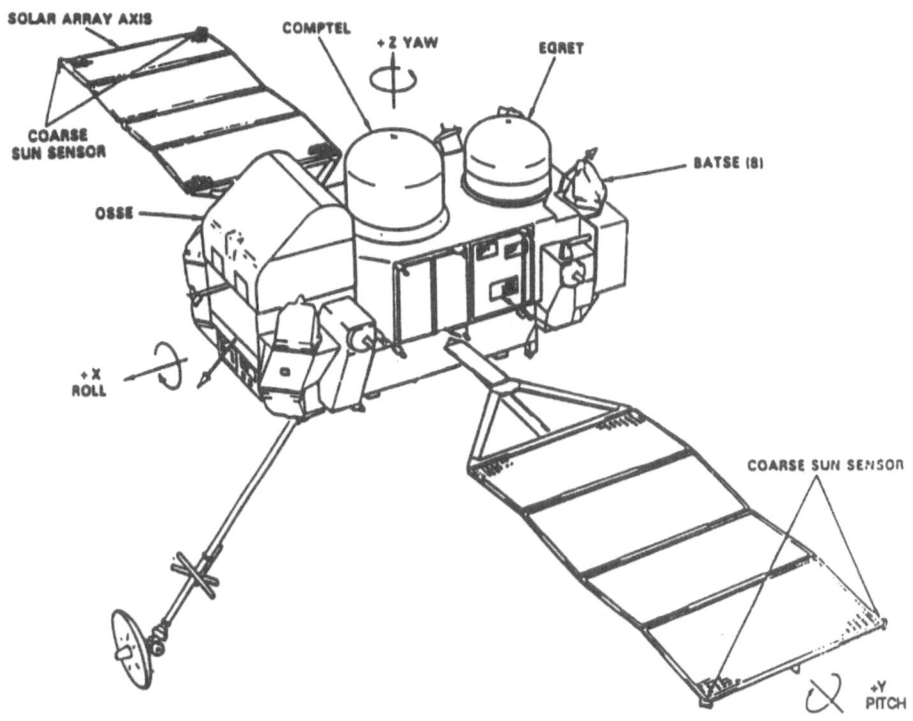

Fig. 8: The Compton Gamma Ray Observatory (GRO)

The configuration of the Compton Observatory is outlined in Figure 8, which illustrates the placement of the four GRO instruments on the spacecraft. Two of the four instruments, COMPTEL and EGRET, view the sky with a wide field of view along the +Z direction of the spacecraft. The OSSE instrument, itself consisting of four independent detector systems, can be independently oriented to point within the spacecraft X-Z plane. The BATSE instrument, consisting of eight modules located on the corners of the spacecraft, is able to view the entire sky.

The four instruments combined weigh approximately 6,000 kg. The observatory can point to any position in the sky and is able to respond to interesting targets of opportunity such as supernovae, novae, and solar flares. While the baseline duration of the Compton Observatory mission is two years, the Observatory contains enough propellant to maintain its orbit for nearly 10 years. A brief description of each of the four Compton Observatory instruments, as described in Kurfess 1992, is given below.

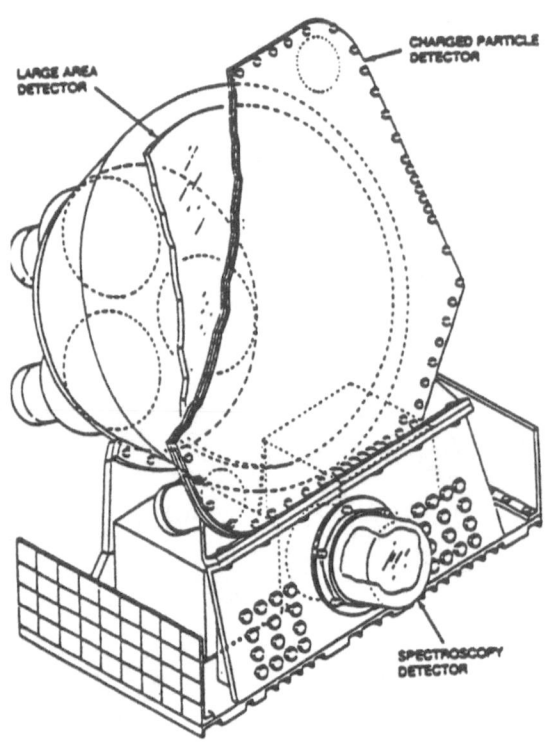

Fig. 9. The Burst and Transient Source Experiment (BATSE) detector module. Eight of the BATSE modules are located on the corners of the Compton Observatory.

The Burst and Transient Source Experiment (BATSE) (Fishman et al. 1989) consists of eight detector modules positioned to provide full sky coverage (note that approximately half the sky is occulted by the earth at any given time). As illustrated in Figure 9, Each BATSE module contains two uncollimated scintillation detectors: a large area detector (LAD), utilizing a 20"-diameter by 0.5"-thick NaI(Tl) crystal which provides high sensitivity and directional capability in the 50 keV - 1 MeV energy region, and a 5.0"-diameter by 3.0"-thick NaI(Tl) spectroscopy crystal which provides better energy resolution and covers the entire range from 10 keV to 100 MeV. The primary objectives of BATSE are to investigate the spatial, temporal, and spectral characteristics of gamma-ray bursts with very high sensitivity. Because of its large area, BATSE can also monitor solar flare activity and can monitor celestial hard X-ray sources during earth occultations. Transient sources with an intensity of about 10% of the Crab nebula flux in the 25-100 keV energy region can be detected on a daily basis using the earth occultation method, enabling BATSE to provide a full sky monitor for bright transients. During normal operations, BATSE accumulates 16-channel spectra with 2-second resolution from its detector systems. 256-channel resolution spectra are acquired from each detector every minute. Upon the detection of a transient event, up to 4 Mbytes of dedicated data can be stored for subsequent transmission to the ground. This can include, for example, 192 high-resolution spectra from the spectroscopy detectors with time resolutions as short as 64 milliseconds. Higher time resolution can be obtained in an event-by-event mode which can store up to 64K events with a time resolution of 128 microseconds.

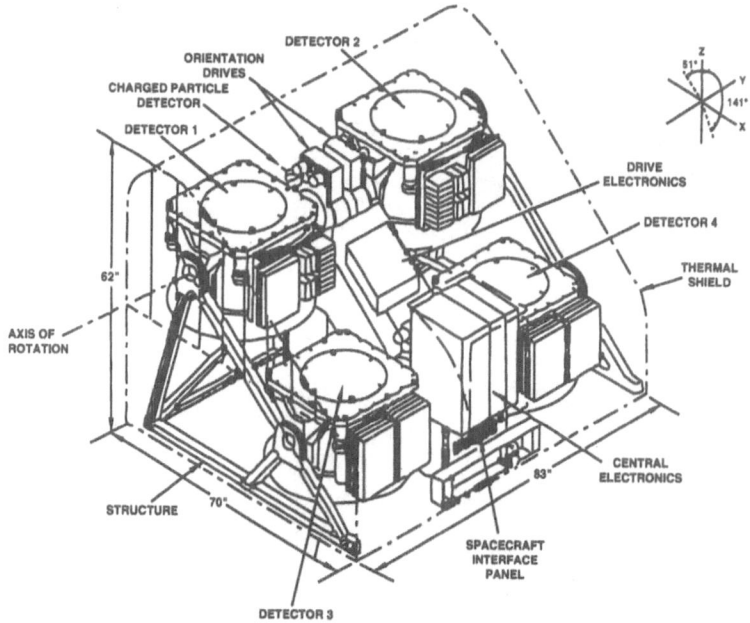

Fig. 10. The Oriented Scintillation Spectrometer Experiment(OSSE).

The BATSE burst trigger algorithm can determine the start of a transient event with a minimum timescale of 64 milliseconds.

Following detection of a transient, BATSE provides a burst trigger signal to the other instruments on the Compton Observatory, which can use this signal to reconfigure their operating modes for acquisition of burst data.

The Oriented Scintillation Spectrometer Experiment (OSSE) (Johnson 1989), shown in Figure 10, provides high sensitivity in the nuclear line region of the spectrum. OSSE consists of four identical detector systems (Johnson et al. 1989), each of which can be rotated independently about the spacecraft y-axis. OSSE has the ability to point its detectors toward sources near the spacecraft X-axis when the spacecraft Z-axis target is occulted by the earth. The operation of the OSSE instrument is controlled by redundant on-board microprocessors. Varied data formats are employed to optimize the spectral acquisitions for the objectives of interest. In response to a BATSE burst signal indicating detection of a solar flare, the four detectors can automatically re-orient to the sun if the sun is located near the X-Y plane. OSSE scientific objectives are chiefly concerned with those phenomena which are particularly associated with nuclear reactions, and include the study of supernovae, novae, solar flares, compact objects, and extragalactic sources.

The Imaging Compton Telescope (COMPTEL) (Ryan 1989) is designed

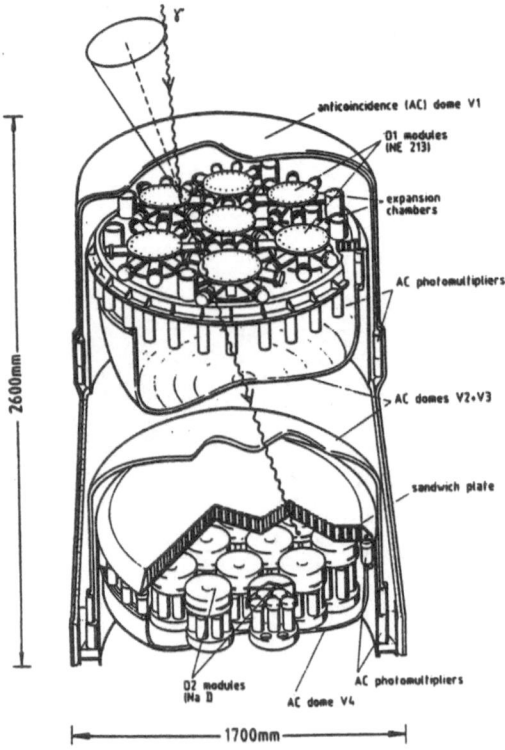

Fig. 11. The Imaging Compton Telescope (COMPTEL).

to observe gamma rays in the 1-30 MeV energy range in a field-of-view
of approximately 1 steradian centered on the spacecraft Z-axis. The
instrument consists of two detector arrays: an upper array, consisting
of seven 28-cm diameter NE-213A liquid scintillation detectors, and a
lower array, consisting of fourteen 28-cm diameter by 3" thick NaI(Tl)
scintillation detectors. An incoming gamma ray Compton scatters in
the upper detector array and the scattered photon is detected in the
lower detector array. In those events where the scattered photon is
totally absorbed in the lower array, the energy of the incident gamma
ray is the sum of the energy losses in the upper and lower detectors,
and the arrival direction is restricted to a cone on the sky whose
axis is determined by the locations of the interactions in the upper
and lower detectors and the Compton scattering angle:

$$\cos(a) = 1 - mc^2(1/E' - 1/E)$$

where a is the Compton scattering angle in the upper detector,
 E is the energy of the incident gamma ray,
 E' is the energy of the scattered gamma ray, and
 mc^2 is the rest energy of the electron.

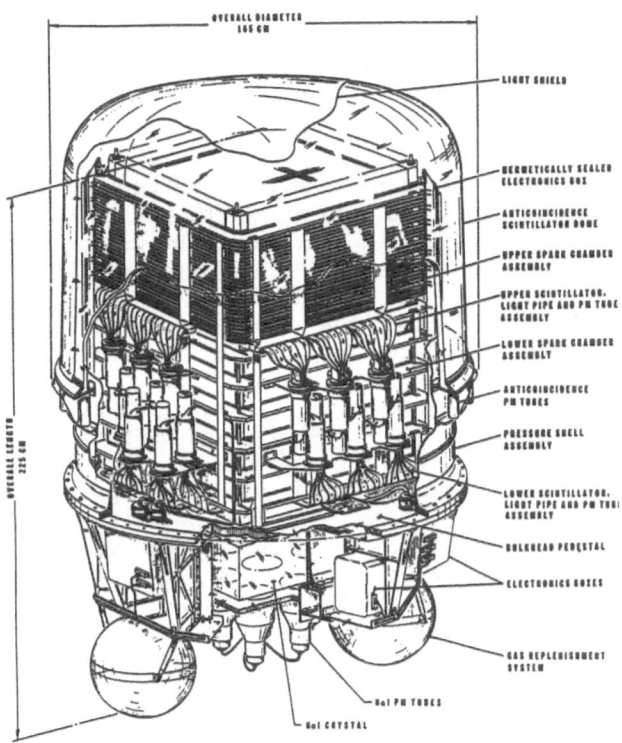

Fig. 12. The Energetic Gamma Ray Experiment Telescope (EGRET).

Using various image reconstruction techniques, COMPTEL can provide images of the gamma-ray sky with a position resolution of several degrees in the 1-30 MeV energy band. During the first 15 months of the Compton Observatory mission, COMPTEL's goals included mapping the entire sky in its energy range.

The Energetic Gamma Ray Experiment Telescope (EGRET) (Kanbach et al. 1989), shown in Figure 12, covers the high-energy portion of the spectrum from 20 MeV to 30 GeV. The instrument consists of two spark chamber modules with tantalum converters which convert incoming gamma rays to positron-electron pairs. The particle trajectories are determined by using a wire readout of the tracks in the spark chambers. The arrival directions of incident gamma rays can be determined to several degrees at 100 MeV and to less than one degree at 1 GeV. EGRET's field-of-view is approximately 45 degrees by 51 degrees (FWHM) centered on the +Z axis. Beneath the spark chamber is a 30" x 30" x 8" thick NaI(Tl) crystal assembly which serves as a total absorption shower calorimeter (TASC) and provides good energy resolution for gamma rays extending into the several GeV region. EGRET will significantly extend the studies initiated by previous missions (SAS-2 and COS-B), improving the sensitivity-effective area product by a factor of 20. During the first 15 months of the Compton Observatory mission, EGRET's goals included mapping the entire sky in its energy range.

9. Explosive Nucleosynthesis

The Solar Maximum Mission (SMM), launched by NASA in 1980 to study solar flare activity and operating for nearly 10 years, carried a NaI(Tl) gamma-ray spectrometer that was able to detect ^{56}Co emission from SN1987A, the Type-2 supernova in the Large Magellanic Cloud (Matz et al. 1988), providing the first direct evidence for explosive nucleosynthesis and confirming the basic ideas regarding models for heavy element synthesis. Recently, the OSSE instrument of the Compton Observatory found evidence for gamma-ray line and continuum emission from ^{57}Co at the expected energy of 122 keV (see Figure 13) (Kurfess et al. 1992 and Leising et al. 1992). The OSSE results provide the first direct measurement of the amount of ^{56}Co produced in the supernova explosion. When compared with contemporaneous bolometric estimates of the luminosity for SN1987A, the OSSE observations imply that ^{57}Co radioactivity does not account for most of the current luminosity of the supernova remnant in low-optical depth models (models having photoelectric absorption losses of 122 keV photons no greater than several percent). Alternate models to account for the OSSE results are proposed in Clayton et al. 1992.

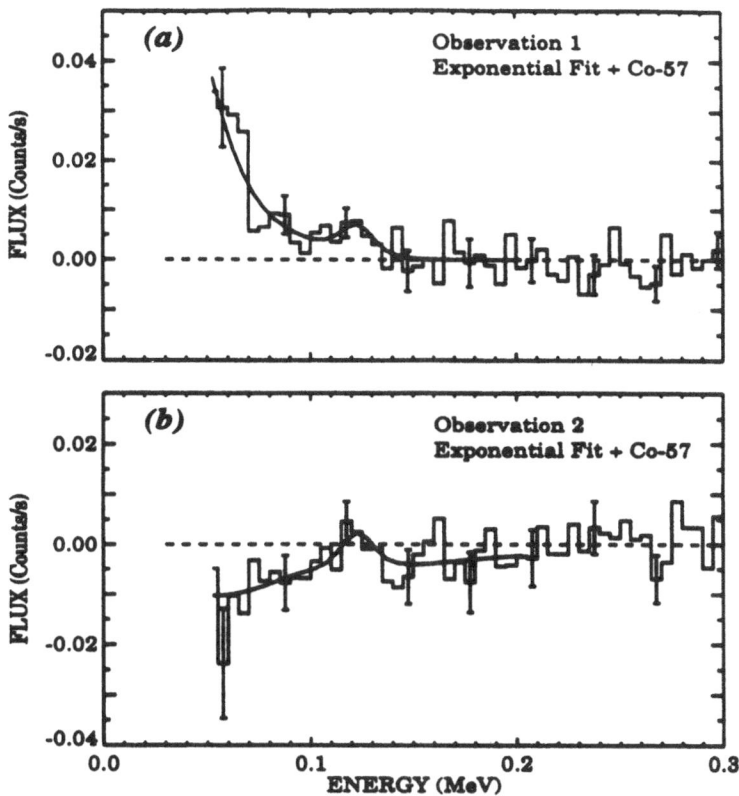

Fig. 13. (a) Energy spectrum for the first OSSE observation of the SN1987A region (1991 July 25 - August 8: 1613-1627 days after the explosion). The solid curve is the best fit for an exponential plus a model 10HMM ^{57}Co template. (b) Energy spectrum for the second OSSE observation of the SN1987A region (1991 December 27 - 1992 January 10: 1768-1782 days after the explosion). The best fit exponential plus model 10HMM ^{57}Co template is shown. The negative exponential component probably reflects an LMC X-3 contribution during background pointings. (Kurfess et al. 1992 and Leising et al. 1992).

HEAO-3 was able to detect 1.809 MeV gamma-ray line emission from the galactic center region (Mahoney et al. 1982) emitted by radioactive ^{26}Al following the isotope's production in massive stars or explosive objects. This was the first non-solar detection of gamma-ray line emission of astrophysical origin. COMPTEL should be able to provide a detailed map of ^{26}Al emission as part of its full sky survey. The half-life of ^{26}Al is approximately 10^6 yr.

10. Gamma Ray Pulsars

Before the Compton Gamma Ray Observatory was launched, only two of the more than 500 radio pulsars were known to emit pulsed gamma radiation: the Crab pulsar (PSR0531+21, 33 ms pulse period), and the Vela pulsar (PSR0833-45, 89 ms pulse period). The Crab, a Type 1 supernova remnant, was a common target for early balloon observations and remains a standard candle for other measurements today due to its stability and emission over a broad band of electromagnetic spectrum, including radio, optical, X-ray and gamma-ray. Vela is an intense source above 70 MeV, but exhibits very weak low-energy gamma-ray and X-ray emission (as mentioned in Kurfess 1992, this might be caused by a large cross section for electron-positron pair creation which would absorb the lower energy radiation produced in the source region).

Fig. 14 shows the average Crab pulsar light curve measured by OSSE (Ulmer et al. 1991) from 1991 May 17-30 in the energy range 60-246 keV. Fig. 15 shows a comparison of the Crab pulsar spectrum measured by OSSE with spectra measured by previous instruments.

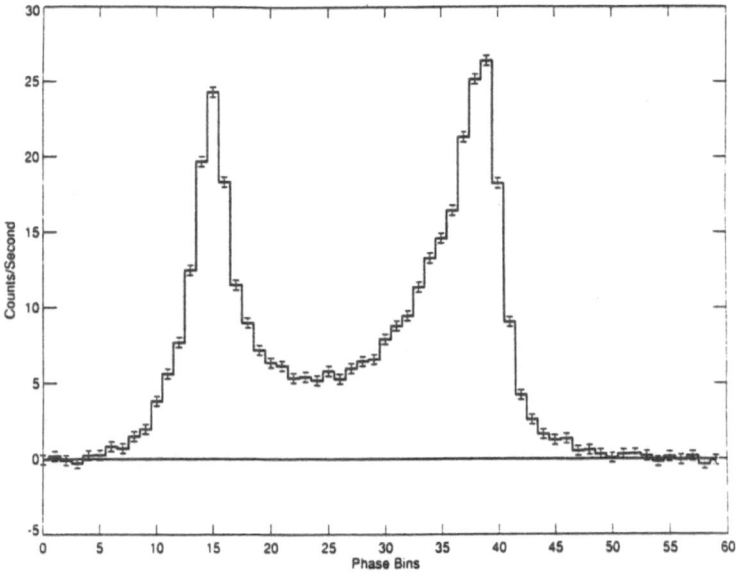

Fig. 14. The average Crab pulsar light curve measured by OSSE from 1991 May 17-30 in the energy range 60-246 keV (Ulmer et al. 1991).

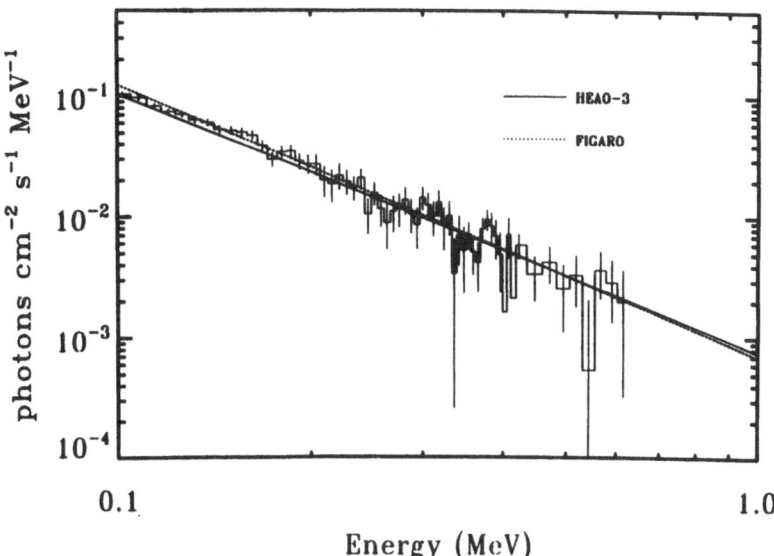

Fig. 15. The average Crab pulsar spectrum measured by OSSE from 1991 May 17-30 (Ulmer et al. 1991), compared with previous spectra measured by HEAO-3 (Mahoney et al. 1984) and FIGARO (Agrinier et al. 1990).

OSSE was able to make the first detection of low-energy gamma-ray emission from the Vela pulsar in the 0.06-0.57 MeV band from measurements made in 1991 August-September and 1992 April-May (Strickman et al. 1992). No significant variability was observed between the two observations, with the light curve having a peak structure similar to that observed at higher energies (Fig. 16). The spectrum measured by OSSE is hard at lower energies and appears to require a break in the 0.5-2 MeV region (Fig. 17).

At least two additional radio pulsars have been observed by the Compton Observatory to emit pulsed gamma radiation: PSR1509-58 (150 ms pulse period) was detected by BATSE at gamma-ray energies from 20 keV to 2 MeV (Wilson et al. 1992) and PSR1706-44 (102 ms pulse period) was detected by EGRET between 100 and 5000 MeV (Thompson et al. 1992). In addition, EGRET has found the high-energy source Geminga (2CG195+04) to have period emission characteristic of an isolated pulsar with a pulse period of 237 ms at energies greater than 50 MeV (Bertsch et al. 1992). The light curve for Geminga, as measured by EGRET, is shown in Fig. 18.

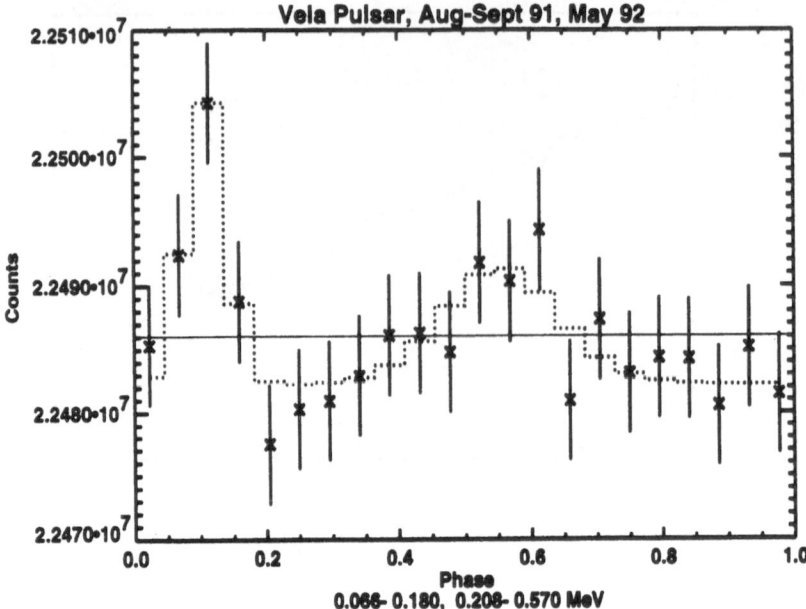

Fig. 16. Vela pulsar light curve measured by OSSE during 1991 August-September and 1992 April-May (Strickman et al. 1992). The energy range is a sum of two bands with high OSSE sensitivity: 0.066-0.180 MeV + 0.208-0.570 MeV. The dotted line is the best fit model and the solid line is the mean.

The mechanism for the gamma-ray emission of these pulsars is not fully understood. As pointed out in Thompson et al. 1992, the gamma-ray emission of these pulsars accounts for as much as 1 percent of their total spin-down energy, a far larger share than is emitted at optical or radio frequencies. Study of the gamma-ray emission from such pulsars is therefore very important in understanding the nature of pulsar emission mechanisms and pulsar evolution.

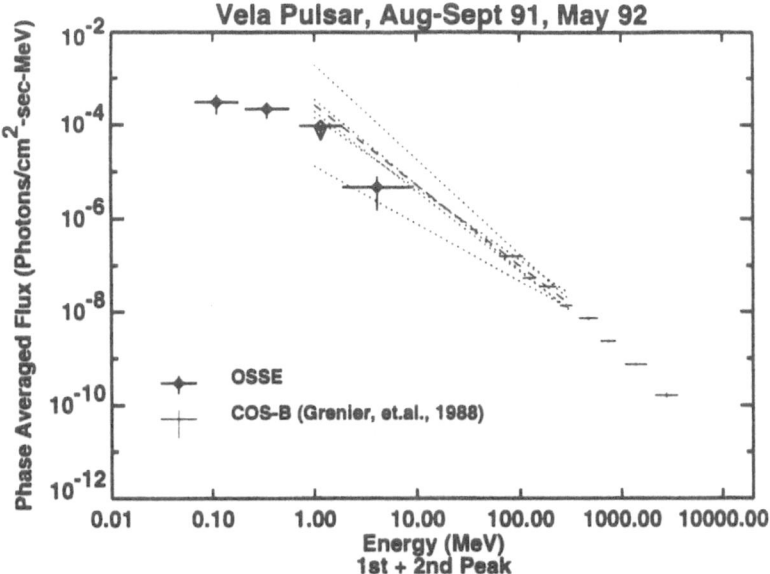

Fig. 17. Vela pulsar spectrum measured by OSSE during 1991 August-September 1991 and 1992 April-May (Strickman et al. 1992). The Flux is the sum of both peaks, averaged over the entire light curve. The high energy data and model spectra are taken from Grenier et al. 1988.

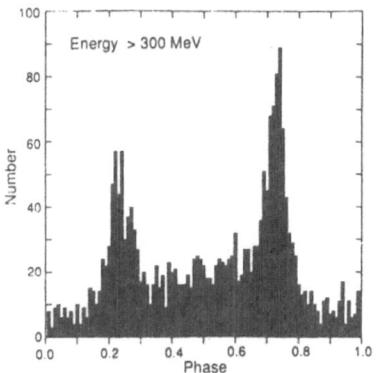

Fig. 18. Light curve of gamma-rays (> 300 MeV) from Geminga measured by EGRET for the combined observation periods 1991 22 April-7 May, 16-30 May and 8-15 June (Bertsch et al. 1992).

11. **Black Hole Candidates**

Early balloon-borne experiments identified hard X-ray emission
from many galactic X-ray sources, many of which are associated with
binary X-ray systems involving a compact object such as a neutron star
or possible black hole. One of the most studied binary X-ray systems
is Cygnus X-1, a black hole candidate which demonstrates variability
on time scales from months to milliseconds.   Figure 19 shows recent
observations from OSSE of the two Galactic black hole candidates,
Cygnus X-1 and GX 339-4, which was observed as a target of opportunity
by OSSE in 1991 September due to a dramatic increase in the flux level
from GX 339-4. One of the observational properties of suspected black
holes is an extremely hard spectrum, out to several keV (Grabelsky et
al. 1992). The relatively good fits of the Sunyaev-Titarchuk model to
the OSSE spectra of Cyg X-1 and GX 339-4 suggest that the spectra are
shaped by Comptonization, despite the deficiency of the fits toward
higher energies.

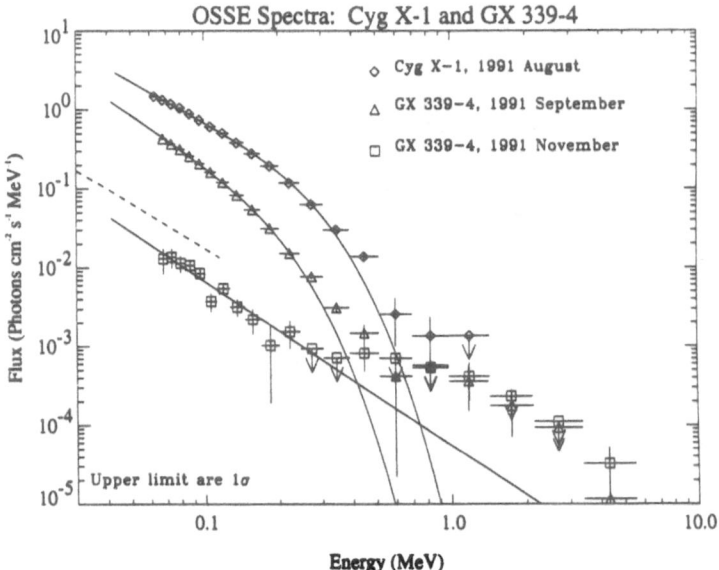

Fig. 19. OSSE spectra of GX 339-4 and Cyg X-1 (Grabelsky et al.
1992). The Cyg X-1 (1991 August) and bright GX 339-4 spectra were
fitted with the Comptonisation model of Sunyaev and Titarchuk
(1980); the dim GX 339-4 spectrum was fitted with a single power
law model.   The solid lines show the best-fit model photon
spectrum; the data points represent the model-dependent
deconvolved count spectra.   The dashed line shows the
extrapolation of the power law portion of the soft X-ray spectrum
of GX 339-4 observed by Makishima et al. (1986).

12. The Galactic Center

Balloon observations were first to detect strong position annihilation radiation coming from the central region of our galaxy. This was first reported by Johnson and Haymes (Johnson et al. 1972) based on a balloon flight in 1971, although the energy for the feature was uncertain. In 1977, Leventhal and co-workers (Leventhal et al. 1978) observed the galactic center with a germanium detector and confirmed the Rice results, and further clearly identified the emission as a narrow 0.511 MeV feature. Thus began the field of positron annihilation astrophysics which as since made for considerable controversy and excitement, and promises to be extremely fruitful in the future.

The initial analysis of data from the HEAO-3 spectrometer, based on two observations of the galactic center region separated by 6 months in 1979-1980 (Reigler et al. 1981), suggested a time-variable, compact source. Subsequent analysis (Mahoney 1988) of the data, however, does not require the source to be time-variable and thus does not require the source to be compact.

These observations have been extended by the OSSE instrument of the Compton Observatory. Figure 20 shows the history of the 511 keV

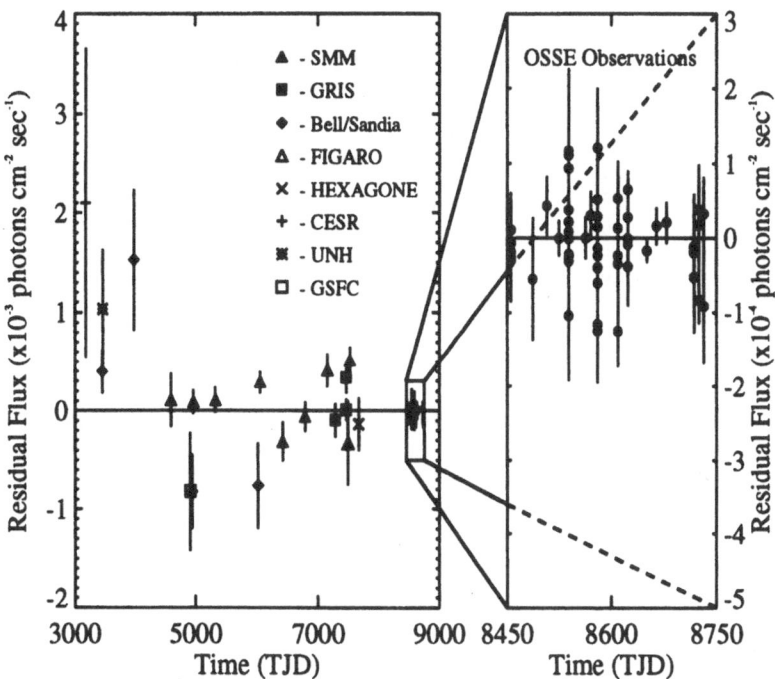

Fig. 20. History of the 511 keV line flux from the galactic center region as observed by various balloon and satellite instruments, including OSSE. (Purcell et al. 1992 and references therein).

line flux from the galactic center region as observed by various balloon and satellite instruments, including the recent OSSE results; OSSE has detected no significant time variability of the line flux. Figure 21 shows the OSSE galactic center spectrum for the period 1991 July 13-24. These observations show conclusive evidence for a narrow 511 keV line and positronium continuum.

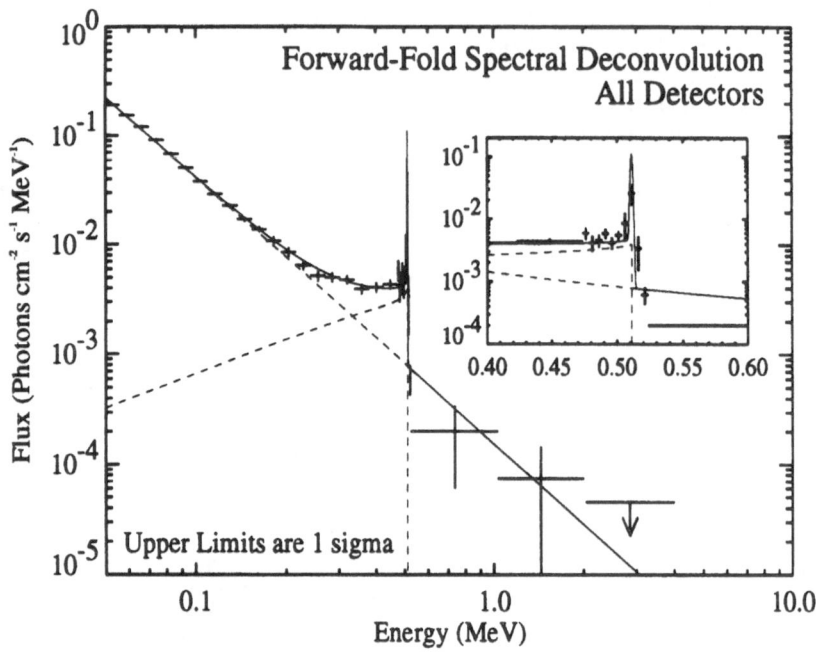

Fig. 21. The OSSE galactic center spectrum for the period 1991 July 13-24. (Purcell et al. 1992 and references therein).

As discussed in Purcell et al. 1992, the OSSE data suggest that the distribution of the 511 keV line emission is composed of two components: 1) a component which is concentrated near the galactic center and is approximately symmetric in galactic longitude and latitude with a FWHM of about 10 degrees and 2) a galactic disk component producing significant emission at longitudes up to approximately 40 degrees, as shown in Figure 22.

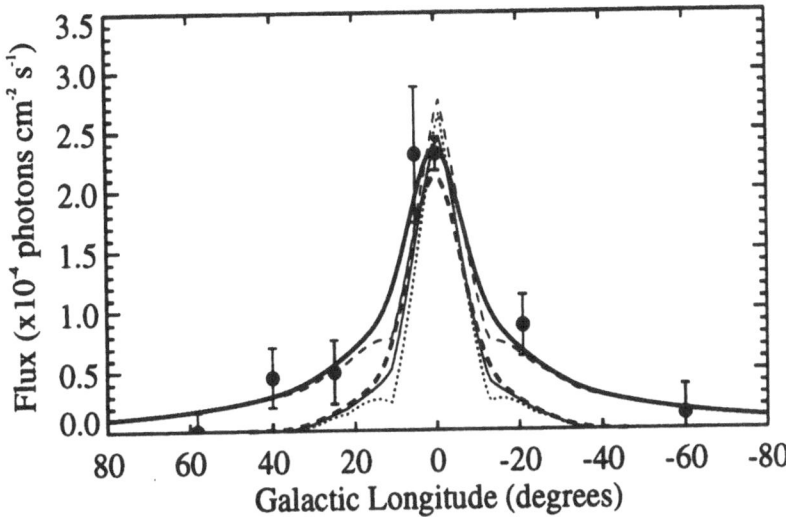

Fig. 22. Distribution of 511 keV flux from galactic center measured by OSSE (Purcell et al. 1992 and references therein).

In October 1990, the French experiment SIGMA (Mandrou 1984) on the Soviet GRANAT satellite was able to detect dramatically increased 300-500 keV emission from the source 1E1740-292 (Sunyaev et al. 1991). This source, initially discovered by the Einstein X-ray Observatory and previously observed to be one of the few hard X-ray sources in the galactic center region (Cook et al. 1991) is about 1 degree from the galactic center and may be associated with a stellar-sized black hole similar to Cygnus X-1.

13. Gamma Ray Bursts

The Vela series of satellites, developed to monitor upper atmosphere nuclear tests, carried gamma ray instruments which first detected and determined the astrophysical origin of gamma-ray bursts (Klebesadel et al. 1973). These are extremely intense bursts of low-energy gamma rays which vary in duration from hundreds of seconds to several milliseconds. Figure 23 shows the variety of time profiles for gamma-ray bursts seen recently by the BATSE instrument of the Compton Observatory (Fishman et al. 1993). The shortest burst seen by BATSE to date had a total duration of less than 5 ms, the shortest ever observed, and contained a significant spike having a width of 0.2 ms (Bhat et al. 1992). Models of these bursts must be able to account for their diversity of shape, structure and duration. The burst time scales seem to require a neutron star origin, suggesting a galactic population of bursts. However, BATSE has recently determined that the angular distribution of bursts is isotropic within its statistical

limits while, at the same time, the number versus intensity
distribution of bursts fails to follow the -3/2 power law expected for
a spatially extended homogeneous distribution of sources (Fig. 24)
(Meegan et al. 1992). This discovery that the bursts appear to be
distributed isotropically but not homogeneously suggests that the
bursts are either distributed in an extended galactic halo or are
cosmological; no known galactic objects have this kind of
distribution. Nearby extragalactic models seem to be excluded because
of a lack of correlations with M31 and the Virgo cluster in the BATSE
data.

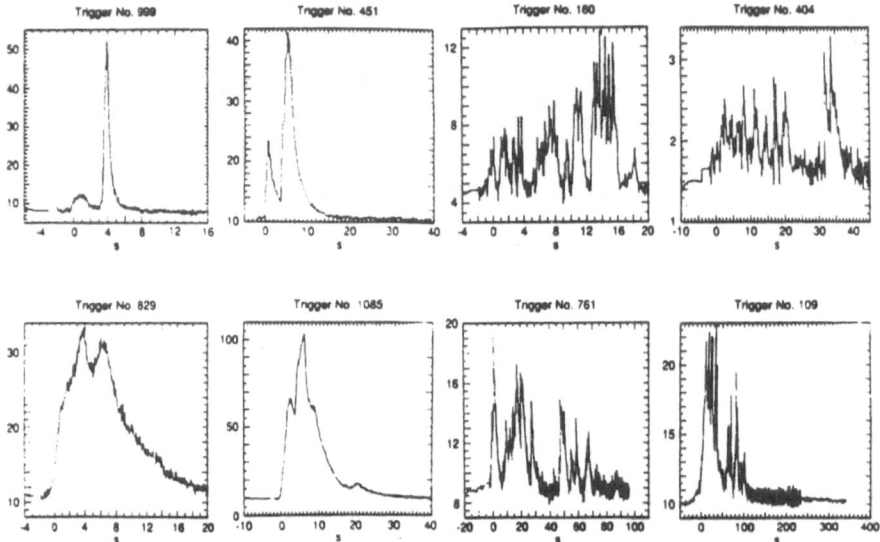

Figure 23. Four gamma-ray bursts with smooth peaks and no apparent
fine time structure are shown in the left two columns. Four
gamma-ray bursts with more complex time profiles are shown in the
right two columns. The energy rangy for all plots is 60-300 keV.
The y-axis units are 1000 counts/s. The x-axis units are seconds
of time. (Fishman et al. 1993).

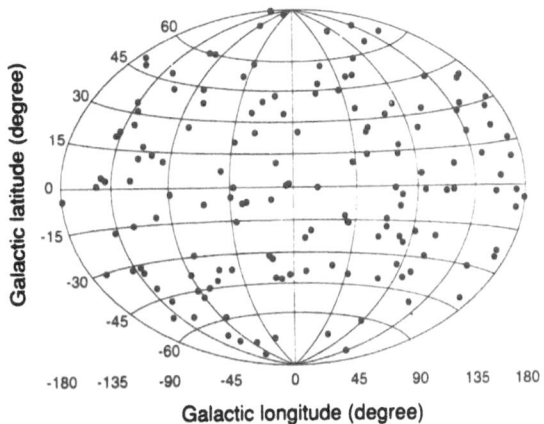

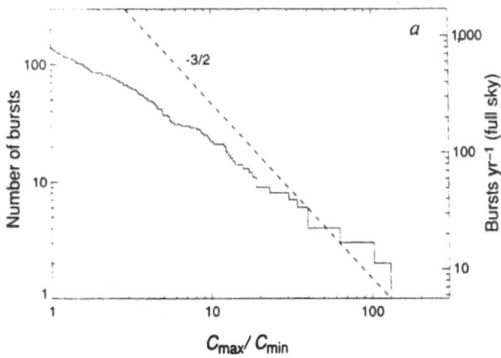

Figure 24. The angular distribution of 153 gamma-ray bursts in galactic coordinates is shown on top. No statistically significant deviation from isotropy is observed. The integral number distribution of 140 gamma-ray bursts as a function of peak rate is shown at bottom. A -3/2 power law is expected for a homogeneous distribution of sources. (Meegan et al. 1992).

14. Gamma Ray Emitting Active Galactic Nuclei

Early balloon flights provided evidence for a cosmic hard X-ray background (Kinzer et al. 1978), and a significant contribution for this background may come from active galactic nuclei (AGN) and quasars. Prior to the launch of the Compton Observatory, Active Galactic Nuclei (AGN) were known to have very hard spectra below 100 keV (Rothschild et al. 1983) which suggested that AGN were prolific sources of gamma rays. COS-B had also detected high-energy gamma-ray emission from the nearby quasar 3C273 (Bignami et al. 1981).

Recently, the EGRET instrument of the Compton Observatory detected intense gamma radiation from the quasar 3C279 in the energy range from 30 MeV to over 5 GeV (Hartman et al. 1992) (Figure 25). This quasar, at a redshift of 0.53, appears to have a gamma-ray luminosity of 10^{48} ergs/s, brighter than any quasar observed prior to EGRET's detection and making this object one of the most energetic sources in nature. Figure 26 shows the luminosity spectra of 3C279 and another quasar, 3C273 (redshift=0.158), showing COMPTEL, EGRET and other data (Hermsen et al. 1992, Hartman et al. 1992, and references therein). In addition to 3C279, EGRET has detected at least 15 other AGN emitting high-energy gamma rays. These observations will provide critical information about the central energy source in Active Galactic Nuclei.

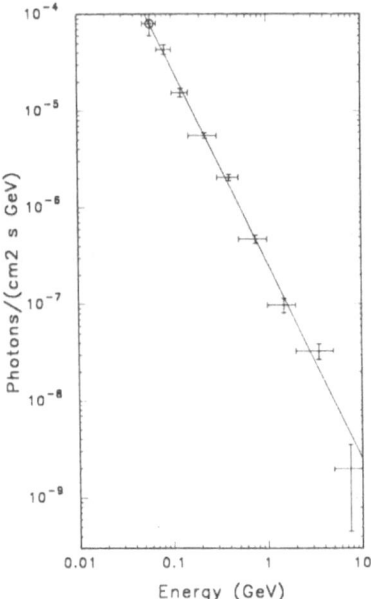

Fig. 25. High-energy gamma-ray spectrum observed for 3C279 by EGRET during the period 1991 June 15-28 (Hartman et al. 1992).

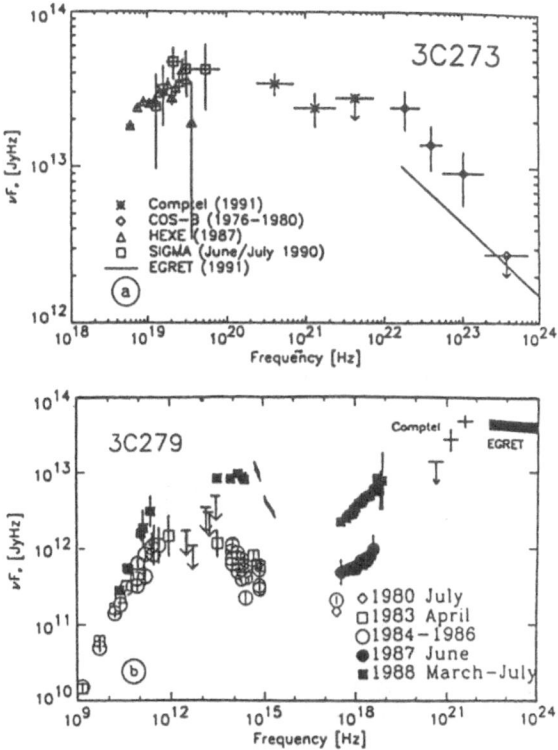

Fig. 26. Luminosity spectra of 3C273 and 3C279 showing COMPTEL, EGRET and other data (see Hermsen et al. 1992, Hartman et al. 1992, and references therein).

15. Future Instruments

The first generation of instruments used for low-energy gamma-ray observations employed broad fields-of-view, typically 15 degrees (FWHM) or more, and provided no imaging capability. In some regions of the sky, however, source confusion is a potential problem for these instruments--most notably in the galactic center region and along the spiral arms of the galaxy. Thus, it became important to build detector systems with imaging capabilities and these have been developed in the last few years by several groups (Althouse et al. 1985, McConnel et al. 1987). The first imaging low-energy gamma-ray experiment to be carried on a satellite was developed in France (Mandrou 1984) and launched on the Soviet GRANAT satellite in December 1989. This experiment, called SIGMA, uses both a position-sensitive scintillation detector and a coded-aperture mask that together provide an instrument which has a 5 degree field-of-view with about 0.1 degree angular resolution. For strong sources, SIGMA can locate a source to within approximately 1 arc minute.

Following the Compton Observatory mission, the next major gamma ray observatory can be expected to combine imaging capability with high spectral resolution in the low-energy gamma-ray range (50 keV - 10 MeV). As described by Kurfess 1992, one approach to achieve this would be to couple coded apertures with high spectral resolution detectors having good position sensitivity (possibly germanium solid state detectors, liquid/solid noble element devices, large-volume silicon devices, or superconducting detectors). An alternative approach would be to use position-sensitive high-resolution detectors in a Compton telescope configuration.

ACKNOWLEDGEMENTS. R.S. would like to thank ONR and C.H.S. thanks NASA for financial support. Both of us thank the organizers M.M. Shapiro, J. Wefel and E. Majorana Center for their hospitality.

REFERENCES
Agrinier, B., et al., 1990, Ap. J., 355, 645.
Althouse, W.E., Cook, W.R., Cummings, A.C., Finger, M.H., Prince, T.A., Schindler, S.M., Starr, C.H., Stone, E.C., 1985, Proc. 19th Intr. Cosmic Ray Conf. (La Jolla), 3, 299.
Berezinsky, V.S. 1976 Proc. 1976 DUMAND Workshop, Univ. Hawaii p. 215.
Berezinsky, V.S., Castagnoli, C. and Galeotti, P. 1986 Ap.J. 301, 235.
Berezinsky, V.S. and Prilutsky, O.F. 1976 Proc. Internat. Conf. Neutrino-76.
Bertsch, D.L., Brazier, K.T.S., Fichtel, C.E., Hartman, R.C., Hunter, S.D., Kanbach, G., Kniffen, D.A., Kwok, P.W., Lin, Y.C., Mattox, J.R., Mayer-Hasselwander, H.A., von Montigny, C., Michelson, P.F., Nolan, P.L., Pinkau, K., Rothermel, H., Schneid, E., Sommer, M., Sreekumar, P. and Thompson, D.J., 1992, Nature, 357, 306.
Bhat, P.N., Fishman, G.J., Meegan, C.A., Wilson, R.B., Brock, M.N. and Paciesas, W.S., 1992, Nature, 359, 217.
Biermann, P.L. and Strittmatter, P.A. 1987, Ap.J. 322, 43.
Bignami, G.F., Bennett, K., Buccheri, R., Caraveo, P.A., Hermsen, W., Sacco, G.B., Scarsi, L., Swanenburg, B.N., Wills, R.D., 1981, Astron. and Astrophys., 93, 71.
Bionta, R.M. et al. 1987 Phys. Rev. Letters 58, 1494.
Blandford, R.O. and Rees, M.J. 1992 Testing the AGN Paradigm p.3 eds, S.S. Holt, S.G. Neff and C. Megan Urry, publ. Am. Inst. Phys. N.Y.
Cameron, R.A., Grove, J.E., Johnson, W.N., Kurfess, J.D., Kinzer, R.L., Kroeger, R.A., Strickman, M.S., Maisack, M., Starr, C.H., Jung, G.V., Grabelsky, D.A., Purcell, W.R., Ulmer, M.P., "OSSE Observations of Active Galaxies and Quasars", Proceedings of the

Compton Centennial Symposium, St. Louis, October 1992, in publication.

Cheng, A. Ruderman, M. and Sutherland, P. 1976 Ap.J. 203, 209.

Clayton, D.D., Leising, M.D., The, L.S., Johnson, W.N., Kurfess, J.D., 1992, Ap. J. (Letters), 399, L141.

Collin-Souffrin, S. 1992 Testing the AGN Paradigm p. 119, ed. S.S. Holt et al. publ. Am. Inst. Phys. N.Y.

Cook, W.R., Grunsfeld, J.M., Heindl, W.A., Palmer, D.M., Prince, T.A., Schindler, S.M., Starr, C.H., Stone, E.C., 1991, Adv. Space Res., 11, 191.

Davis, R., Jr., Evans, J.C., Cleveland, B.T. 1978, Neutrinos-78, p. 53, ed. E.C. Fowler, publ. Purdue University.

Dermer, C.D., "Gamma Rays from Active Galactic Nuclei", Proceedings of the Compton Centennial Symposium, St. Louis, October 1992, in publication.

Eichler, D. 1981 Proc. 1980 DUMAND Symposium, 2, 266.

Fabian, A.C. 1992, Testing the AGN Paradigm, p. 657, ed. S.S. Holt et al., publ. Am. Inst. Phys., N.Y.

Fichtel, C.E. et al. 1992, Bull. Am. Phys. Soc. 37, 997.

Fichtel, C.E., "Some Aspects of the Scientific Significance of High Energy Gamma Ray Astrophysics", 1991, Adv. Space Res., 11, (8)303.

Fishman, G.J., Meegan, C.A., Wilson, R.B., Paciesas, W.S., Parnell, T.A., Austin, R.W., Regage, J.R., Matteson, J.L., Teegarden, B.J., Cline, T.L., Schaefer, B.E., Pendleton, G.N., Berry Jr., F.A., Horack, J.M., Storey, S.D., Brock, M.N., Lestrade, J.P., 1989, Proceedings of the First GRO Science Workshop, 2-39.

Fishman, G.J., Meegan, C.A., Wilson, R.B., Paciesas, W.S., Pendleton, G.N., Harmon, B.A., Horack, J.M., Brock, M.N., Kouveliotou, C. and Finger, M., "Overview of Observations from BATSE on the Compton Observatory", 1993, Astron. Astrophys. Suppl. Ser., in press.

Ginzburg, V. and Syrovatskii, S.I. 1964 The Origin of Cosmic Rays, MacMillan, N.Y.

Grabelsky, D.A., Matz, S.M., Purcell, W.R., Ulmer, M.P., Johnson, W.N., Kinzer, R.L., Kroeger, R.A., Kurfess, J.D., Strickman, M.S., Grove, J.E., Cameron, R.A., Jung, G.V., "OSSE Spectral Observations of GX339-4 and Cyg X-1", Proceedings of the Compton Centennial Symposium, St. Louis, October 1992, in publication.

Grenier, I.A., Hermsen, W., Clear, J., 1988, Astron. Astrophys., 204, 117.

Gunn, J.E. and Ostriker, J.P. 1969 Nature 221, 454.

Hartman, R.C., Bertsch, D.L., Fichtel, C.E., Hunter, S.D., Kanbach, G., Kniffen, D.A., Kwok, P.W., Lin, Y.C., Mattox, J.R., Mayer-Hasselwander, H.A., Michelson, P.F., von Montigny, C., Nel, H.I., Nolan, P.L., Pinkau, K., Rothermel, H., Schneid, E., Sommer, M., Sreekumar, P. and Thompson, D.J., 1992, Ap. J. (Letters), 385, L1.

Hermsen, W., Aarts, H.J.M., Bennett, K., Bloemen, H., de Boer, H., Collmar, W., Connors, A., Diehl, R., van Dijk, R., den Herder, J.W., Kuiper, L., Lichti, G.G., Lockwood, J.A., Macri, J., McConnell, M., Morris, D., Ryan, J.M., Schonfelder, V., Simpson,

G., Steinle, H., Strong, A.W., Swanenburg, B.N., de Vries, C., Webber, W.R., Williams, O.R., Winkler, C., "COMPTEL detections of the quasars 3C273 and 3C279", 1993, Astron. Astrophys. Suppl. Ser., in press.

Hirata, K. et al. 1987 Phys. Rev. Letters 58, 1490.

Johnson, W.N., Kurfess, J.D., Kinzer, R.L., Purcell, W.R., Strickman, M.S., Jung, G.V., Ulmer, M.P., Jensen, C.M., Share, G.H., Clayton, D.D., Dyer, C.S., Cameron, R.A., 1989, Proceedings of the First GRO Science Workshop, 2-22.

Johnson, W.N., Harnden, F.R., Haymes, R.C., 1972, Ap. J., 172, 61.

Kanbach, G., Bertsch, D.L., Favale, A., Fichtel, C.E., Hartman, R.C., Hofstadter, R., Hughes, E.B., Hunter, S.D., Hughlock, B.W., Kniffen, D.A., Lin, Y.C., Mattox, J.R., Mayer-Hasselwander, A., Montigny, C.V., Nolan, P.L., Pinkau, K., Rothermel, H., Schneid, E., Sommer, M., Thompson, D.J., Walker, A.H., 1989, Proceedings of the First GRO Science Workshop, 2-1.

Kazanas, D. and Ellison, D.C. 1986 Ap.J. 304, 178.

Kinzer, R.L., Johnson, W.N., Kurfess, J.D., 1978, Ap. J., 222, 370.

Klebesadel, R.W., Strong, I.B. and Olson, R.A., 1973, Ap. J. (Letters), 182, L85.

Kurfess, J.D., Johnson, W.N., Kinzer, R.L., Kroeger, R.A, Strickman, M.S., Grove, J.E., Leising, M.D., Clayton, D.D., Grabelsky, D.A., Purcell, W.R., Ulmer, M.P., Cameron, R.A., Jung, G.V., "OSSE Observations of 57Co in SN1987A", 1992, Ap. J. (Letters), 399, L137.

Kurfess, J.D., "Gamma Ray Astrophysics: Development and Future Prospects", 1992, Annals of the New York Academy of Sciences, 655, 292.

Leising, M.D., Clayton, D.D., The, L.S., Johnson, W.N., Kurfess, J.D., Kinzer, R.L., Kroeger, R.A., Strickman, M.S., Grove, J.E., Grabelsky, D.A., Purcell, W.R., Ulmer, M.P., Cameron, R.A., Jung, G.V., "Compton Observatory OSSE Studies of Supernovae and Novae," Proceedings of the Compton Centennial Symposium, St. Louis, October 1992, in publication.

Leventhal, M., MacCallum, C.J., Stang, C.J., 1978, Ap. J. (Letters), 225, L11.

Leiter, D. and Kafatos, M. 1978 Ap. J. 226, 32.

Lovelace, R.V.E. 1976 Nature 262, 649.

Mahoney, W.A., Ling, J.C., Jacobson, A.S., Tapphorn, R.M., 1980, Nucl. Instr. Methods, 178, 363.

Mahoney, W.A., Ling, J.C., Jacobson, A.S., Lingenfelter, R.E. 1982, Ap. J., 262, 742.

Mahoney, W.A., Ling, J.C., Jacobson, A.S., 1984, Ap. J., 278, 784.

Mahoney, W.A., 1988, AIP Conf. Proc., 170, 149.

Maisack, M., Johnson, W.N., Kinzer, R.L., Strickman, M.S., Kurfess, J.D., Jung, G.V., Grabelsky, D.A., Purcell, W.R., Ulmer, M.P., "OSSE Observations of NGC 4151," Proceedings of the Compton Centennial Symposium, St. Louis, October 1992, in publication.

Makino, F. et. al. 1989 Ap.J. (Letters) 347, L9.

Makishima, K., et al., 1986, Ap. J., 308, 635.

Manchester, R.N. and Taylor, J.H. 1977 Pulsars, publ. by Freeman, San Francisco.

Mandrou, P., 1984, Adv. Space Res., 3, 524.

Marscher, A.P. and Brown, R.L. 1978 Ap. J. 220, 474.

Matz, S.M. et al. 1988 Nuclear Spectroscopy of Astrophysical Sources, p. 51, eds. N. Gehrels and G.H. Share, AIP Conf. Proc. 170.

Maraschi, L. Ghisellini, G. and Celotti, A. 1992 Testing the AGN Paradigm, p. 439 ed. S.S. Holt et al., publ. Am. Inst. Phys. N.Y.

Matteson, J.L., 1978, Proc. AIAA, 78, 35.

Matz, S.M., Share, G.H., Leising, M.D., Chupp, E.L., Vestrand, W.T., Purcell, W.R., Strickman, M.S., and Reppin, C., 1988, Nature, 331, 416.

McConnell, M.L., Dunphy, P.P., Forrest, D.J., Chupp, E.L., Owens, A., 1987, Ap. J., 321, 543.

Meegan, C.A., Fishman, G.J., Wilson, R.B., Paciesas, W.S., Pendleton, G.N., Horack, J.M., Brock, M.N. and Kouveliotou, C, 1992, Nature, 355, 143.

Morisawa, K. and Takahara, F. 1989 Publ. Astron. Soc. Japan 41, 873.

Murphy, R.J., Share, G.H., Grove, J.E., Johnson, W.N., Kinzer, R.L., Kroeger, R.A., Kurfess, J.D., Strickman, M.S., Matz, S.M., Grabelsky, D.A., Purcell, W.R., Ulmer, M.P., Cameron, R.A., Jung, G.V., Jensen, C.M., Vestrand, W.T., Forrest, D.J., "OSSE Observations of Solar Flares", Proceedings of the Compton Centennial Symposium, St. Louis, October 1992, in publication.

Protheroe, R.J. and Kazanas, D. 1983 Nature, 302, 228.

Purcell, W.R., Grabelsky, D.A., Ulmer, M.P., Johnson, W.N., Kinzer, R.L., Kurfess, J.D., Strickman, M.S., Jung, G.V., "OSSE Observations of Galactic 511 keV Annihilation Radiation", Proceedings of the Compton Centennial Symposium, St. Louis, October 1992, in publication.

Ramana Murthy, P.V. and Wolfendale, A.W. 1986 Gamma Ray Astronomy, Cambridge Univ. Press, Cambridge.

Reigler, G.R., Ling, J.C., Mahoney, W.A., Wheaton, W.A., Willett, J.B., Jacobson, A.S., Prince, T.A., 1981, Ap. J. (Letters), 248, L13.

Rothschild, R.E., Mushotzky, R.F., Baity, W.A., Gruber, D.E., Matteson, J.L., Peterson, L.E., 1983, Ap. J., 269, 423.

Ryan, J.M., 1989, Proceedings of the First GRO Science Workshop, 2-11.

Schonfelder, V., Aarts, H.J.M., Bennett, K., Bloemen, H., de Boer, H., Busetta, M., Collmar, W., Connors, A., Diehl, R., den Herder, J.W., Hermsen, W., Kuiper, L., Lichti, G.G., Lockwood, J.A., Macri, J., McConnell, M., Morris, D., Much, R., Ryan, J., Simpson, G., Stacy, J.G., Steinle, H., Strong, A.W., Swanenburg, B.N., Taylor, B.G., Varendorff, M., de Vries, C., Webber, W., Winkler, C., "An Overview of First Results from COMPTEL", 1993, Astron. Astrophys. Suppl. Ser., in press.

Shapiro, M.M. and Silberberg, R. 1979a Proc. Internat. Conf. Cosmic Rays (Kyoto) 10, 363.

Shapiro, M.M. and Silberberg, R. 1979b Proc. Internat. Conf. Cosmic Rays (Kyoto) 10, 352.

Silberberg, R. and Shapiro, M.M. 1979 Proc. Internat. Cosmic Ray
    Conf. (Kyoto) 10, 357.
Stanev, T. 1992 High Energy Neutrino Astrophysics Workshop, ed. V.J.
    Stenger et al. Univ. of Hawaii, World Scientific Publ. Co. to be
    publ.
Stecker, F.W. et al. 1991 Phys. Rev. Letters 66, 2697.
Strickman, M.S., Grove, J.E., Johnson, W.N., Kinzer, R.L., Kroeger,
    R.A., Kurfess, J.D., Grabelsky, D.A., Matz, S.M., Purcell, W.R.,
    Ulmer, M.P., "OSSE Observations of the Vela Pulsar," Proceedings
    of the Compton Centennial Symposium, St. Louis, October 1992, in
    publication.
Sunyaev, R.A., Titarchuk, L.G., 1980, Astron. Astrophys., 86, 121.
Sunyaev, R., Churazov, E., Gilfanov, M., Pavlinsky, M., Grebenev,
    S., Dekhanov, I., Kuznetsov, A., Yamburenko, N., Ballet, J.,
    Laurent, P.L., Paul, J., Salotti, L., Natalucce, L., Niel, M.,
    Roques, J.P., Mandrou, P., 1991, Proc. Intr. Symposium on
    Gamma-Ray Line Astrophysics (Paris), 29.
Szabo, A.P. and Protheroe, R.J. 1992 High Energy Neutrino
    Astrophysics Workshop ed. V.J. Stener et al., Univ. Hawaii,
    World Scientific Publ. Co., to be publ.
Thompson, D.J., Arzoumanian, Z., Bertsch, D.L., Brazier, K.T.S.,
    D'Amico, N., Fichtel, C.E., Fierro, J.M., Hartman, R.C., Hunter,
    S.D., Johnson, S., Kanbach, G., Kaspi, V.M., Kniffen, Lin, Y.C.,
    Lyne, A.G., Manchester, R.N., Mattox, J.R., Mayer-Hasselwander,
    H.A., Michelson, P.F., von Montigny, C., Nel, H.I., Nice, D.,
    Nolan, P.L., Pinkau, K., Rothermel, H., Schneid, E., Sommer, M.,
    Sreekumar, P. and Taylor, J.H., 1992, Nature, 359, 615.
White, R.S. and Silberberg, R. 1991 Cosmic Rays, Supernovae and the
    Interstellar Medium, p. 213, ed. M.M. Shapiro et al. Kluwer Acad.
    Publ., Dordrecht.
Wilson, R.B., et al., 1992, IAU Circ., No. 5429.
Ulmer, M.P., Matz, S.M., Cameron, R.A., Grabelsky, D.A., Grove,
    J.E., Johnson, W.N., Jung, G.V., Kinzer, R.L., Kurfess, J.D.,
    Leising, M.D., Purcell, W.R., Strickman, M.S., "OSSE Observations
    of the Crab Pulsar," 1991, Proceedings of the Second GRO Science
    Workshop, 253.

POINT SOURCES OF TEV AND PEV COSMIC GAMMA RAYS

E. J. FENYVES
University of Texas at Dallas
P.O. Box 830688
Richardson, Texas 75083-0688

ABSTRACT. The study of PEV gamma ray sources seems to be the best method to search for the origin of high energy cosmic ray nuclei. Detection of TeV and PeV gamma ray sources is the only method which combined with other observations, such as anisotropy measurements, can answer the question whether the ultra high energy cosmic ray particles are accelerated in point sources, or on large scales. The two major experimental methods used for observing and measuring very high energy cosmic gamma rays, the atmospheric Cerenkov technique and the extensive air shower technique are reviewed. With the rapid development of very high energy gamma ray astronomy a relatively large number of TeV and PeV gamma ray sources has been observed using the above techniques, but only one TeV source, the Crab Nebula is completely well confirmed, and is the best candidate for the standard candle of TeV gamma ray astronomy. Observations of other potential TeV and PeV gamma ray point sources are presented, and some of the major questions of very high energy gamma ray astronomy and astrophysics are discussed.

1. Introduction

1.1. ORIGIN OF HIGH ENERGY COSMIC RAYS

The search for the origin of high energy cosmic rays is one of the most challenging problems in the long history of cosmic ray research. Tremendous efforts, both experimental and theoretical have been made to resolve this problem, but in spite of many decades of research carried out by a rather large number of authors this problem is still open and unresolved. The major avenues to yield important information about the sources of cosmic ray particles above 1 TeV are studies of small anisotropies in their flux, the spectrum and mass composition of these particles, and, particularly, the detection of point sources of TeV and PeV cosmic gamma rays [1, 2].

Among these methods the observation of PeV gamma ray sources seems to be the most direct approach to study the origin of high energy cosmic ray particles because - according to our present knowledge - these gamma rays can arise only from π° decay, i.e., from interactions of cosmic ray nuclei well above 1 PeV [3]. In contrast to this, TeV gamma ray emission may be explained also through synchrotron or curvature radiation of electrons (or positrons). Thus PeV gamma ray emitters could be readily identified as sources of high energy cosmic ray nuclei.

1.2. OBSERVATION OF COSMIC RAY POINT SOURCES

Charged cosmic ray particles cannot be used for detecting cosmic ray point sources because the charged particles are deflected in the interstellar magnetic field. Even 1 PeV single charged primary particles have a Larmor radius of about 0.3 pc ($\sim 1.10^{18}$ cm), and have their initial directions almost completely isotropised.

M. M. Shapiro et al. (eds.), Particle Astrophysics and Cosmology, 95–110.

Among the neutral primary particles neutrons are unstable, and have a mean life of 889 sec which corresponds even for a relatively small mean range of 1 kpc to $E_n > 10^{17}$ eV energy. Neutrinos have very small interaction cross sections, and according to this only extremely strong sources such as supernovas, e.g., SN 1987A can be detected by their neutrino emission. Until now no steady neutrino emitting point source has been observed.

Thus the only neutral primary cosmic ray particles which can be used for the observation of cosmic ray point sources are gamma rays, and - as shown above - the study of PeV gamma ray sources seems to be the most important method to search for the origin of high energy cosmic ray nuclei.

2. Gamma Ray Astronomy and Astrophysics above 1 TeV: Experimental Methods

2.1. EXPERIMENTAL METHODS

The rapid development of very high energy gamma ray astronomy was essentially based on two detection methods:

(1) The atmospheric Cerenkov technique for energies from 100 GeV to 100 TeV

(2) The extensive air shower technique for energies above 100 TeV.

These ground-based detection methods are using large detector arrays, particularly for the observation of extensive air showers, enabling them to observe very small gamma ray intensities. The space-based gamma ray telescopes measuring smaller energy gamma rays are, in contrast much smaller. The space-based and ground - based detector technologies give an overlap around 100 GeV.

2.2. THE ATMOSPHERIC CERENKOV TECHNIQUE

A high energy gamma ray incident on the top of the atmosphere initiates an electromagnetic cascade shower. The relativistic electrons and positrons in the shower with velocities above the local light velocity in the atmosphere produce Cerenkov light which is detected by the atmospheric Cerenkov detector. The density of Cerenkov light near the core is approximately proportional to the primary gamma energy.

The simple atmospheric Cerenkov detector consists of

(a) a mirror and

(b) a photomultiplier tube with aperture that defines the acceptance angle of the detector,

(c) and fast, pulse counting electronics (Fig. 1).

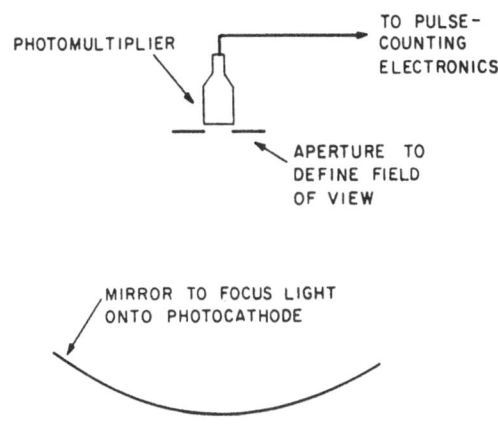

Figure 1. The essential elements of an atmospheric Cerenkov detector [1].

The high energy gamma ray initiated electromagnetic cascade showers are profoundly different in their development from the proton generated air showers in their (Fig. 2)

(a) lateral distribution
(b) angular distribution
(c) composition
(d) spectral distribution, and
(e) temporal distribution.

The major developments to improve the sensitivity and performance of the atmospheric Cerenkov detectors were concentrated in the last decades around the following efforts:

(a) reduction of the threshold energy
(b) improvement of the angular resolution
(c) reduction of background by primary identification
(d) increase of the number of detectors.

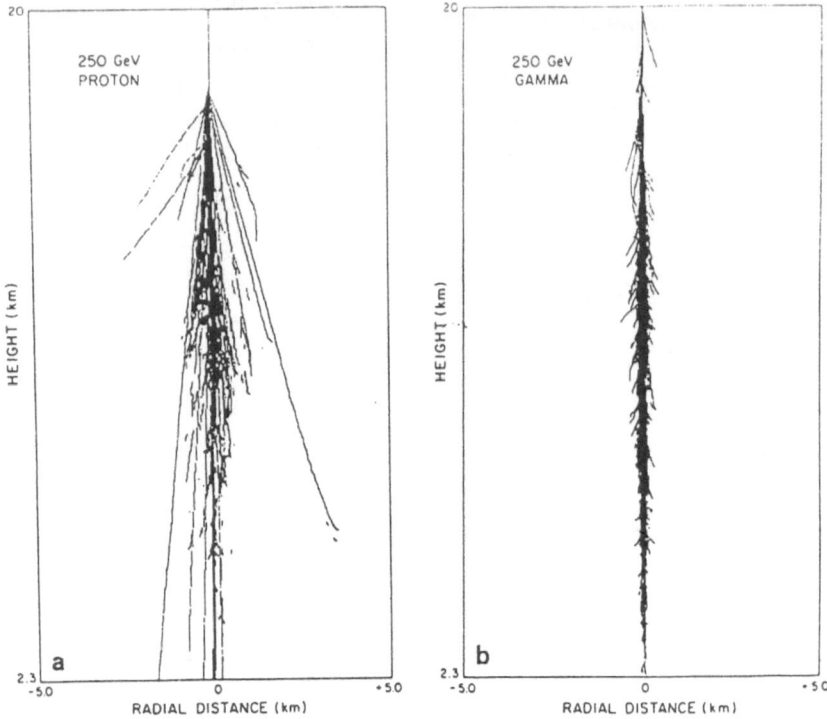

Figure 2. Development of proton and gamma-ray initiated air showers in the atmosphere [1].

The threshold energy can be reduced by increasing the mirror collection area. For example the construction of the Smithsonian's 10 m Reflector (Fig. 3) lowered the energy threshold form 2 TeV to 0.2 TeV, a very significant achievement in the atmospheric Cerenkov detector technique [1].

The angular resolution has been improved by differential fast timing between separated detectors (typically 100 m apart, see Fig. 4), using smaller apertures, applying imaging techniques with large arrays of photomultipliers in the focal plane (Fig. 3), or multiplexing detectors.

Figure 3. The 10 m Optical reflector at the Whipple Observatory in southern Arizona. In the focal plane there is an array of 109 photomultipliers. The outlying 1.5 m telescopes are used to further characterize the shower [4].

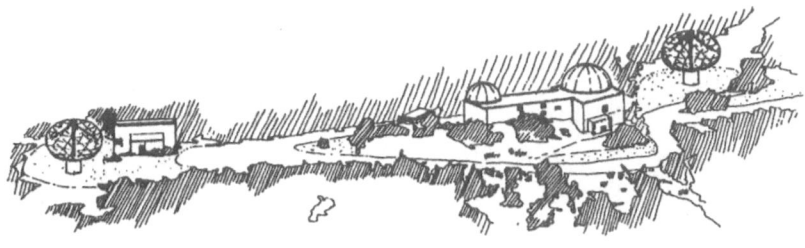

Figure 4. The two atmospheric Cerenkov detectors on the 7,600 foot ridge of Mount Hopkins at a 120 m distance from each other [5].

The background reduction was achieved by using the difference in the lateral distribution, and the time pulse structure of gamma and proton initiated showers, and using the image size differentiation of showers. The latter one being the most successful technique is based mainly on the inherent difference in size of the image because π production angles in proton initiated showers are large as compared to the angles of the purely electromagnetic processes.

A large number of atmospheric Cerenkov detectors are in now operation or under construction (Table 1, and ref. [6] in this volume). Recent developments made in the last few years in improving the measurement techniques are the following:

(a) More sophisticated age selection and better statistical evaluation of the observed date
(b) Development of models predicting the phenomena to be observed
(c) More accurate calibration of the detectors to determine absolute fluxes and energies
(d) Improvement of the Cerenkov imaging technique.

TABLE 1. List of VHE Gamma-ray Observatories [4]

Institution(s)	Observatory	Hemisphere	Birthdate	Description
Potchefstroom	Potchefstroom, South Africa	S	1985 1989	4 (3 x 1.5m) 8 "
Durham	Narrabri, Australia	S	12986	3 x 3m
Adelaide	Alice Springs, Australia	S	1989	3 x 3m
Auckland, Tokyo, Hobart	New Zealand Australia	S S	1987 1989	3 x 2m 3.8m camera
Smithsonian Bartol, Leeds	South Pole	S	1989	2 x 0.5m
Crimea A. O.	Crimea	N	1970 1988	2 (2 x 1.5m) 6 x 4m cameras
Yerevan	Mt. Aragats,	N		4 x 3m cameras
Tata	Pachmari, Armenia, U.S.S.R.	N	1987	12 x 1.5m
Smithsonian, Iowa, Leeds, Dublin, Michigan	Whipple Observ., Arizona	N	1968 1982 1989	10m 10m camera 2 x 10m camera
Wisconsin, Purdue, Columbia, Hawaii, Athens	Haleakala Hawaii	N	1983 1988	6 x 1.5m 12 x 1m
Durham	La Palma,	N	_1988	3 x 3m
Beijing	Beijing	N		
Saclay	Pyrenees	N	1988	

2.3 THE EXTENSIVE AIR SHOWER TECHNIQUE

At primary energies larger than 10^{14} eV the electromagnetic cascade can penetrate through the approximately 28 radiation length deep atmosphere to sea level and produce an extensive air shower (EAS).

The EAS detector arrays are detecting primarily the electron component at the earth surface, i.e., sea level or mountain altitudes. The size of the electron detectors is of the order of ≥ 1 m^2 extending to several hundred meters or more.

The muon component is detected by means of strongly shielded muon detectors, or deep underground muon arrays, e.g., the Homestake Surface-Underground Telescope. The EAS's are in general formed by hadronic primary particles, mainly protons and to a much lesser extent by heavier nuclei. One of the largest and most powerful EAS detectors the Utah-Michigan-Chicago array consisting of a large number of electron detectors, muon detectors and Cerenkov telescopes is shown in Figure 5.

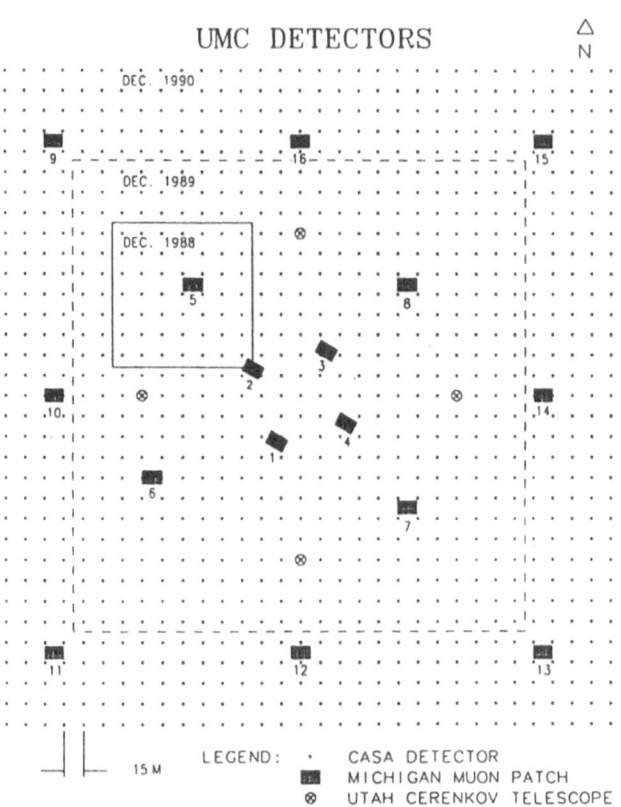

Figure 5. The Utah-Michigan-Chicago shower array at Dugway, Utah [6]. (CASA stands for the Chicago Air Shower Array.

The energy of the shower, E is calculated from N_e, the total number of electrons by means of the E vs. N_e relationship computed for different altitudes from computer simulations using the known or extrapolated cross sections for strong and electromagnetic processes. The shower size, N_e can be determined by fitting the electron densities measured at the individual electron counters to the lateral distribution of the electron component given by Nishimura and Kamata

$$\rho(r) = \frac{N}{2\pi r_1^2} f(s) \left(\frac{r}{r_1}\right)^{s-2} \left(1 + \frac{r}{r_1}\right)^{s-4.5}$$

where $\rho(r)$ is the electron density per m^2 at a distance of r from the shower core; s the age parameter of the shower; $f(s)$ a function of s, and r_1 is the Moliere scattering unit (~ 80m at sea level), or using other parametrized analytical formulas for the lateral distribution [7].

From the fit the shower size

$$\int_0^\infty 2\pi r \rho(r) dr$$

and the core location and age parameter can be determined.

The arrival direction of the shower is obtained from fast timing information of the electron counters. From this the celestial coordinates of the shower direction can be calculated.

Reduction of the background, i.e., the EAS's generated by hadronic primaries, is carried out by
  (i) determining s, and
  (ii) the muon content of the shower.

Hadronic showers have a nuclear core, and thus a lesser fast attenuation in the atmosphere. In contrast, pure electromagnetic showers are characterized by faster attenuation, and thus are "older" showers. Their age parameter s is larger and their lateral distribution is flatter.

The muons are generated in the decay of charged pions and other mesons, e.g.

$$\pi^\pm \rightarrow \mu^\pm + \nu_\mu \left(\bar{\nu}_\mu\right)$$

The photo-nuclear cross section for generating pions in the pure electromagnetic cascades is very small, about 100 times smaller than the cross section of pion production in nuclear interactions of hadrons. According to this, the muon content of gamma initiated showers is about an order of magnitude smaller than that of hadronic showers.

The major EAS arrays in the 0.1-10 PeV energy range are shown in Table 2.

TABLE 2: Air Shower Experiments (0.1-10 PeV) [8]

Group/Location	Energy Threshold (PeV)	Comment
Baksan, USSR	0.2	
Moscow State University, USSR	1	
Tien Shan (Nikolsky et al.) USSR	0.15	
Mt. Aragatz, USSR		Major new installation
Ooty, India (Tata, Sreekantan, Tonwar)	0.2	No muon detection
KGF, India, Tata	0.5	200 m^2 of surface muon counters
U. of Pennsylvania/UT-Dallas	0.1	Surface-Underground Detector
Akeno, Japan	> 1	Sea level
GREX, Leeds, U.K.	> 0.5	Sea level
Kiel et al. at La Palma		Under construction
Heidelberg tracking air shower experiment		Major new effort/technique (proposal)
Swiss air shower experiment (tracking)		Prototype
Japanese exp. (Suga) at Chacaltaya		
U. of Utah, Fly's Eye		Not a γ-detector per se
Chicago - Michigan - Utah	~0.1	Major new effort - largest area ever in muon detectors
"Cygnus" array at Los Alamos	< 0.1	Expansion in progress
JANZOS in New Zealand	~ 0.1	
SPASE (Bartol/Leeds at South Pole)	~ 0.05	
Buckland Park, Australia	~ 1	Sea Level
Notre Dame (Poirier)		Proposed tracking EAS detector, sea level
Torino - EAS top at Gran Sasso		Construction this summer, surface/underground connection
Karlsruhe		Major new proposal
GRANDE		
RPC (Italian)		New idea to use resistive plate chambers for both upward and downward events

3. Very High Energy Gamma Ray Astronomy: Experimental Results

3.1. TeV GAMMA RAY SOURCES

With the rapid development of very high energy gamma ray astronomy and astrophysics a relatively large number of TeV gamma ray point sources has been observed using the atmospheric Cerenkov technique. A careful analysis carried out by T. Weekes shows, however, that only one of them, the Crab Nebula is completely well confirmed [1,2,9]. In addition to this, five more sources can be considered as potential candidates:

1. Cygnus X-3
2. Hercules X-1
3. Vela X-1
4. 4U0115 + 63
5. Mkn 421

In addition to the small number of confirmations there are a number of problems and peculiarities with these observations which make the selection of very high energy gamma ray sources rather difficult:

(a) The apparent gamma ray fluxes are very similar and not far above the cosmic ray background
(b) The sources have very flat spectra. If the background cosmic rays are the limiting factor at all energies, then the minimum detectable flux must have an integral spectral index of - 0.8. The spectral index measured for Cygnus X-3 is about -1.0
(c) Many of the detected sources are transients and highly variable.

Among the TeV gamma ray emitters the Crab Nebula is the best studied object. The Crab Nebula (Crab Pulsar) is the brightest supernova remnant (SNR) in the Galaxy, and is at about 2 kpc from the earth. It has been observed to show long term steady activity at TeV energies since 1985 with a very high statistical significance (35 σ). It has signals which have properties agreeing in every respect with those expected from a gamma ray source. There is no evidence for variability on any time scale, and there is no periodicity.

The total gamma ray flux of the Crab is 7.10^{-11} photons cm^{-2} s^{-1} above 0.4 TeV, and the energy spectrum between 0.4 and 4 TeV is

$$N(E)dE = 2.5.10^{-10}(E / 0.4TeV)^{-2.4 \pm 0.3}$$

as measured by the Whipple collaboration [10] in good agreement with the result of the Michigan group [11] (see Fig. 6).

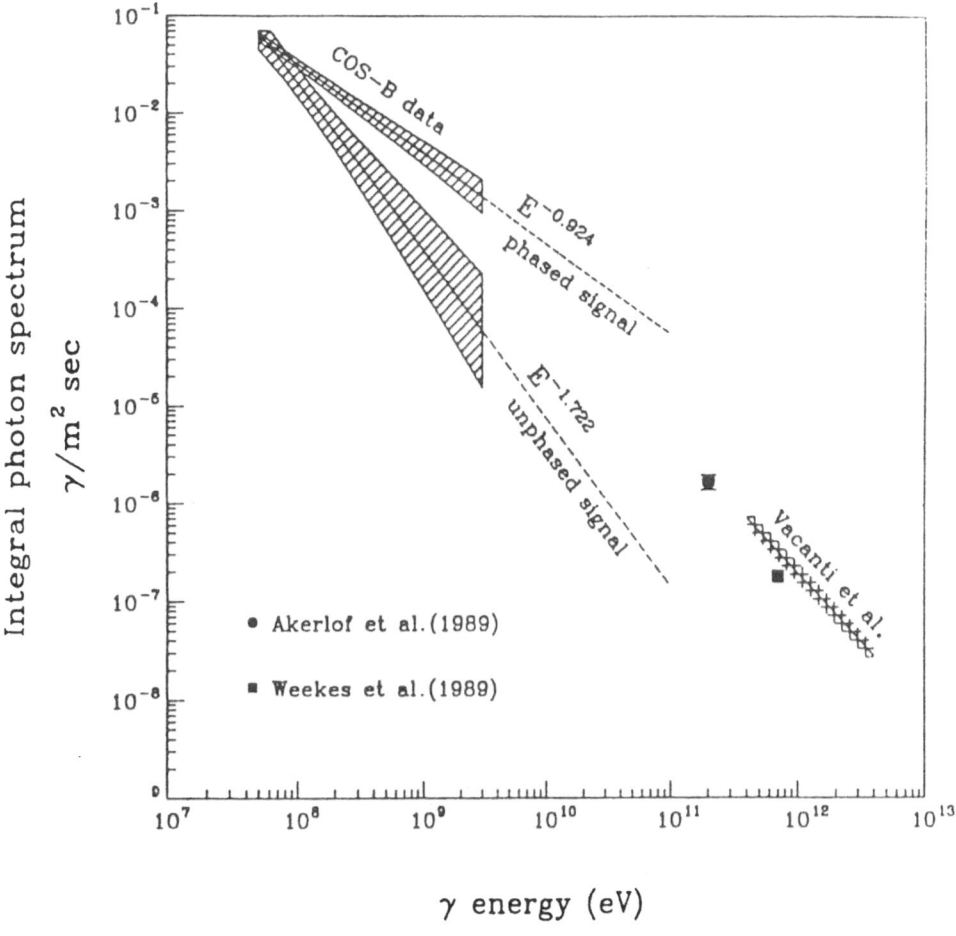

Figure 6. The measured integral fluxes from the Crab Nebula [18].

Among the other TeV candidates the binary X-ray source Cygnus X-3 was reported in more than 15 papers [2] as a gamma ray source from energies of 100 MeV to 0.5 EeV. Cygnus X-3 (> 11 kpc) has a very broad energy range, rather flat energy spectrum (Fig. 7), and a very large variability at all energies [2]. There seems to be a steady decrease in the gamma ray luminosity in the last decade by at least a factor of 10, and recently below the level of detectability. There is some evidence that the periods of gamma ray emission are associated with radio flares, such as the September 1972, September 1980, and the October 1985 radio flares [15].

The detection of a 12.59 ms periodicity by the Durham group [16] was not observed by other groups and seems to be controversial [17].

According to this Cygnus X-3 may have been a point source up to 1 PeV, but it has declined in intensity and is no longer detectable with telescopes of the same sensitivity as those used in the previous experiments.

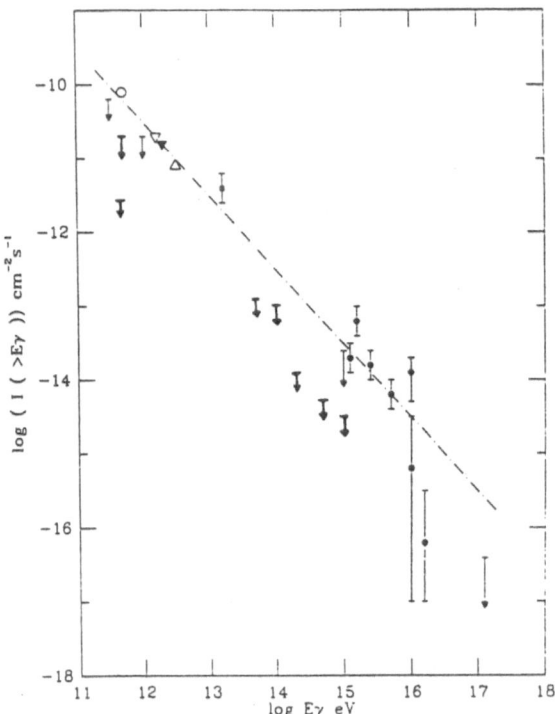

Figure 7. The measured spectrum of Cygnus X-3 [18].

Hercules X-1 another binary X-ray source (5 kpc) has been reported by a number of groups [2] as an always episodic (usually tens of minutes) and periodic (1.2 sec) TeV gamma ray source. In the spring and summer of 1986 three groups, the Haleakala, Whipple and Los Alamos groups obtained the most significant observation of episodic emission from any binary source in the form of three blue-shifted observations with 1.236 sec periodicity [2] (consistent with each other but significantly different from the known X-ray period; see Fig. 8).

The major problem with the Hercules X-1 observations is, however, that they do not show the characteristic features of gamma ray emission. For example the Los Alamos group [19] (Cygnus collaboration) measuring higher energy showers (200 TeV) with their EAS array including a muon detector, obtained a muon-to-electron ratio as high as in the hadronic background events.

According to this more evidence is required before Hercules X-1 could be regarded as an established high energy gamma ray source.

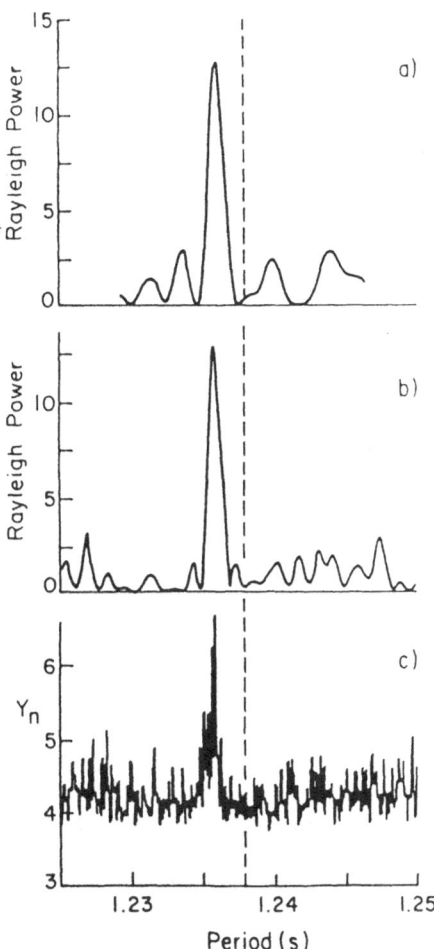

Figure 8. Power as a function of period for three sets of observations of Hercules X-1 in 1986 [2].

Vela X-1 is a high mass binary X-ray source (0.6 kpc) with a very strong magnetic field (10^{13-14} Gauss) in the neutron star, was reported by several groups as a steady periodic source [18]. Those experiments, however, had no capability to distinguish gamma ray showers from the background hadronic showers. According to this there is some strong but not enough convincing evidence for Vela X-1 to be a TeV gamma ray source.

4U0115+63 is a transient X-ray pulsar which was observed as a relatively strong periodic TeV gamma ray emitter by the Durham group [20]. Weak evidence for short episodes was reported by other groups. The Tata group reported, however, no evidence for emission, and the Whipple group withdrew later its confirmation [21]. According to this the conservative conclusion is that 4U0115+63 is not a steady periodic source but it may be a sporadic one.

The BL Lac object Mkn 421 which has been established as an X-ray source long ago, was recently reported as a TeV gamma ray source by the Whipple Observatory collaboration [22]. Mkn 421 is relatively close to the earth as indicated by its small redshift (z = 0.0308), and the above detection supports the suggestion that only nearby Active Galactic Nuclei may be visible in the TeV gamma ray because such emission from distant sources would be strongly attenuated by interactions with extragalactic IR radiation.

In the last few years an additional candidate AE Aquarii, a cataclysmic variable containing a white dwarf and not a neutron star, has been reported by two groups, but further observations are needed to establish this object a high energy gamma ray source[18].

3.2. PeV GAMMA RAY SOURCES

At PeV energies where the gamma ray sources are studied with extensive air shower arrays, the situation is even more confusing. Several groups reported earlier PeV gamma ray emissions from the Crab Nebula, while other groups failed to observe these ultra high energy gamma rays [3, 23, 24].

More recently, however, the Baksan group [23] reported a PeV energy gamma ray burst from the direction of the Crab Nebula on February 23, 1989. This observation was later confirmed by the Tata (KGF) group, the Gran Sasso collaboration and the Tien Shan group [25]. The PeV burst from Crab, lasting for about seven hours, has high statistical significance according to Rao and Sreekantan [26], but is considered only marginally significant by Weekes [27]. The muon component measured by the KGF group showed nearly the same number of muons as in the hadronic background showers.

The conservative conclusion is that there is not enough strong evidence for PeV gamma ray emission from the Crab Nebula.

The initial observation of Cygnus X-3 as a PeV source by the Kiel [28] and Leeds [29] groups with EAS arrays opened the possibility that this is a major source of cosmic rays. However, the Kiel group showed that the muon content of these showers was essentially the same as that of the hadronic background showers.

These and other subsequent positive observations [3, 15] of PeV gamma rays from Cygnus X-3 were, however, not substantiated by recent measurements [30, 31] and it seems that the source has now fallen below its previous level of detectable PeV emission.

The Utah group [32] using the Fly's Eye detector reported recently a significant excess of gamma rya showers at energies > 0.5 EeV. The Haverah Park group [33] did not find any excess of same energy air showers and present a lower upper limit for gammas than the flux observed by the Utah group. The Akeno group [34] reported, however, a significant (3.5σ) excess of > 0.5 EeV air showers from Cygnus X-3 with a similar flux of gammas to that observed by the Utah group.

The above inconsistencies in the observations of ultra high energy gamma rays from Cygnus X-3 resulted in critical reviews of this object as a source of high energy gammas and cosmic ray nuclei [35, 36].

In conclusion, the observations on Cygnus X-3 are contradicting and confusing. At best Cygnus X-3 may have been a PeV gamma ray emitter but has strongly declined in intensity in the past decade.

Hercules X-1 has been observed as a PeV emitter by the Baksan [23] and Tata [37] groups, and by the Cygnus collaboration [19], but recent observations carried out with the Mt. Hopkins array did not substantiate those results [24].

4. Conclusions

As we have shown the question of the origin of high energy cosmic rays approached via the detection of TeV and PeV gamma ray sources is very complex and difficult. It is, however, the only method which combined with other observations, such as anisotropy measurements, can answer the basic question whether the ultra high energy cosmic ray particles above 0.1 PeV are accelerated in point sources or on large scales, e.g., in a galactic wind [39]. The particle acceleration mechanism in compact sources, as well as the problem of "muon rich" showers are some of the major questions to be resolved [40, 41].

It was shown that only one high energy gamma ray source, the Crab Nebula is completely well confirmed and only in the TeV energy range. More observations carried out with larger and significantly improved detectors characterized by increased sensitivity for gamma ray detection and hadron rejection capability are needed. An impressive number of new detectors and experimental arrays have been proposed or are already under construction. There is a good chance that this new generation of experiments will bring us much closer to resolving the problem of the origin of high energy cosmic rays.

5. References

[1]   T.C. Weekes, High Resolution Gamma Ray Cosmology, Nuclear Physics B (Proc. Suppl.) 10B, 41 (1989).
[2]   T. C. Weekes, 14th Texas Symposium on Relativistic Astrophysics, 1988, Annals of the New York Academy of Sciences, Vol. 571, p. 372.
[3]   A.A. Watson, Proc. 19th ICRC, La Jolla, 1985, Vol. 9, p. 111.
[4]   T.C. Weekes, TeV Gamma-Ray Astronomy, Harvard-Smithsonian Center for Astrophysics, Preprint Series No. 3310 (1991).
[5]   C.W. Akerlof et al., Arkansas Gamma-Ray and Neutrino Workshop, Nuclear Physics B (Proc. Suppl.) 14A, 237 (1990).
[6]   R.A. Ong, Arkansas Gamma-Ray and Neutrino Workshop, Nuclear Physics B (Proc. Suppl.) 14A, 273 (1990).
[7]   E. Fenyves et al., Phys. Rev. D37, 649 (1988).
[8]   Report of the HEPAP Subpanel on High Energy Gamma Ray and Neutrino Astronomy, U.S. Department of Energy, Office of Energy Research, 1988.
[9]   T. Weekes, Texas/ESO-CERN Symposium on Relativistic Astrophysics, Cosmology, and Fundamental Physics, Annals of the New York Academy of Sciences, Vol. 647, p. 326 (1991).
[10] G. Vacanti et al., Harvard-Smithsonian Center for Astrophysics, Preprint Series No. 3200 (1990).

[11] C. Akerlof et al., Proc. 21st ICRC, Adelaide, 1990, Vol. 2, p. 135.

[12] M.J. Lang et al., Proc. 21st ICRC, Adelaide, 1990, Vol. 2, p. 139.

[13] O.T. Tumer et al., Proc. 21st ICRC, Adelaide, 1990, Vol. 2, p. 155.

[14] O.T. Tumer et al., Nuclear Physics B (Proc. Suppl.) 14A, 176 (1990).

[15] R.J. Protheroe, Proc. 20th ICRC, Moscow, 1987, Vol. 8. p. 21.

[16] P.M. Chadwick et al., Very High Energy Gamma Ray Astronomy, ed. K.E. Turner, Dordrecht: Reidel, p. 115 (1987).

[17] K. O'Flaherty et al., Proc. 21st ICRC, Adelaide, 1990, OG 4.1-2.

[18] T.C. Weekes, Harvard-Smithsonian Center for Astrophysics, Preprint Series No. 3310 (1991).

[19] B.L. Dingus et al., Phys. Rev. Lett. 61, 1906 (1988).

[20] P.M. Chadwick et al., Astron. Ap. 151, L1 (1985).

[21] D.J. Macomb et al., Harvard-Smithsonian Center for Astrophysics, Preprint Series No. 3192 (1991).

[22] M. Punch et al., Nature 358, 477 (1992).

[23] V.V. Alexeenko et al., Proc. 20th ICRC, Moscow, 1987, Vol. 1, p. 219.

[24] G.H. Gillanders et al., Proc. 21st ICRC, Adelaide, OG 4.1-9.

[25] V.V. Alexeenko et al., Proc. of International Workshop on Very High Energy Gamma Ray Astronomy, eds. A.A. Stepanian, D.J. Fegan and M.F. Cawley) p. 187 (1989).

[26] M.V.S. Rao and B.V. Sreekantan, Current Science, Vol. 62, 617 (1992).

[27] T.C. Weekes, Harvard-Smithsonian Center for Astrophysics, Preprint Series No. 3247 (1991).

[28] M. Samorski and W. Stamm, Proc. 18th ICRC, Bangalore, 1983, Vol. 11, p. 244, and Ap. J. Lett. 268, L17 (1983).

[29] J. Lloyd-Evans et al., Nature 305, 784 (1983).

[30] P.J.V. Eames et al., Proc. 20th ICRC, Moscow, 1987, Vol. 1, p. 210.

[31] V.V. Alexeenko et al., Proc. 20th ICRC, Moscow, 1987, Vol. 1, p. 229.

[32] G.L. Cassiday et al., Phys Rev. Lett. 62, 383 (1989).

[33] M.A. Lawrence et al., Phys. Rev. Lett. 63, 1121 (1989).

[34] M. Teshima et al., Phys. Rev. Lett. 64, 1628 (1990).

[35] J.M. Bonnet-Bidaud and G. Chardin, Physics Reports 170, Vol. 6, p. 326 (1988).

[36] G. Chardin and G. Gerbier, Astron. Ap. 210, 52 (1989).

[37] B.S. Acharya et al., Arkansas Gamma-Ray and Neutrino Workshop, Nuclear Physics B (Proc. Suppl) 14A, 216 (1990).

[38] Yu A. Fomin et al., Proc. 20th ICRC, Moscow, 1987, Vol. 1, p. 397.

[39] J.R. Jokipii and G.E. Morfill, Ap. J. 290, L1 (1985), and 312, 170 (1987).

[40] A.K. Harding, Arkansas Gamma Ray and Neutrino Workshop, Nuclear Physics B (Proc. Suppl.) 14A, 3 (1990).

[41] T.K. Gaisser, Arkansas Gamma Ray and Neutrino Workshop,, Nuclear Physics B (Proc. Suppl.) 14A, 381 (1990).

A SEARCH FOR RADIATIVE NEUTRINO
DECAY FROM SUPERNOVAE

R.S. Miller and R.C. Svoboda
Department of Physics and Astronomy
Louisiana State University
Baton Rouge, LA 70803-4001

Presented by Richard S. Miller

Abstract

If a massive neutrino species exists then it is possible that it has a radiative decay mode. Cosmological arguments require that a neutrino of mass $> 100eV$ be unstable. Motivated by these considerations, a 3-D model of radiative neutrino decay has been developed and used to simulate the decay evolution of supernova neutrinos. A sensitive search for the decay gamma rays using extragalactic supernovae and SN1987a as neutrino sources is proceeding using the COMPTEL instrument aboard the Compton Gamma Ray Observatory. The non-observation of decay gamma rays would place more stringent mass/lifetime limits on the neutrino than current limits by about 1-6 orders of magnitude (depending on mass). Such a systematic analysis has never before been attempted. In addition, the production of the decay gamma rays over cosmological time-scales is being studied to determine its possible relationship to the diffuse gamma ray emission.

I. The Search for Massive Neutrinos

The search for the existence of a massive neutrino species is an area of intense theoretical and experimental research. Until recently, laboratory studies indicated that the observed properties of the neutrino were consistent with zero rest mass. The notable exceptions have been the experiments studying solar neutrinos and nuclear beta decay.

For over 20 years the study of electron neutrino production in our Sun has led to the so called "Solar Neutrino Problem". So far, all solar neutrino experiments have detected a lower flux than predicted [1]. One compelling explanation for this

111

M. M. Shapiro et al. (eds.), Particle Astrophysics and Cosmology, 111–118.
© 1993 *Kluwer Academic Publishers.*

is the introduction of neutrino flavor oscillations, which require neutrinos to possess mass [2].

Recently, evidence has been building for the existence of a heavy 17-keV neutrino. This neutrino has been indirectly observed by studying the electron spectra from the beta decay of a number of nuclear isotopes [3]. The deviation in the observed endpoint energies of the beta spectra can be interpreted as the existence of a 17-keV neutrino mixing at the 1% level with a "normal" massless ν_e. Accelerator limits on $\nu_\mu \to \nu_e$ oscillations indicate that the massive neutrino is not associated with ν_μ [4]. Since the measurement of the Z^o width appears to allow only three types of neutrinos with mass less than 45 GeV, this massive 17-keV neutrino must be the ν_τ. Although the evidence is controversial and awaits unambiguous confirmation the current results make the search for massive neutrinos even more intriguing.

There are also theoretical reasons to believe that neutrinos have mass. Massless neutrinos are not required by the Standard Model of particle physics. In fact it has been necessary to introduce the conserved quantum number of "lepton flavor" to distinguish the three types of neutrinos. This internal degree of freedom might manifest itself in mass differences, as it does for the charged leptons.

Very massive neutrinos ($> 100eV$) are expected to be unstable. Astrophysical limits based on the requirement that the mass density of the universe not exceed $\Omega = 1$ require $\Sigma m_\nu < 100eV$ [5]. Thus a 17-kev ν_τ should decay. A very reasonable decay mode to expect would then be:

$$\nu_\tau \to \nu_{e,\mu} + \gamma \qquad (1)$$

This mode is the simplest two-body decay that does not violate any known conservation laws (except lepton flavor). No new particles are required and angular momentum sum rules are satisfied.

II. A Search for Radiative Neutrino Decay

We are currently performing a sensitive search for radiative neutrino decay in conjuction with the COMPTEL instrument team. The recently begun search uses type-II supernovae as the source of neutrinos. The high angular resolution and sensitivity of the COMPTEL gamma ray telescope [6] provides a unique oportunity to study any supernova gamma ray emissions. By observing SN1987A and extragalactic supernovae, any observed emissions can be analyzed and the relevant neutrino mass and lifetime computed (assuming a radiative decay mode) as described in the following sections. The non-observation of gamma rays from these sources will place significant limits on neutrino mass and lifetime.

III. Supernova Neutrinos

The detection of neutrinos from SN1987A [7] confirmed the central theoretical prediction of neutrino production in type-II supernovae: approximately 3×10^{53} ergs of binding energy released primarily as neutrinos on a time scale of a few seconds. The neutrinos detected by the IMB and Kamiokande collaborations were very likely $\bar{\nu}_e$, based on the relevant neutrino-water cross sections. It is widely believed, however, that since the $\bar{\nu}_e$'s were probably produced by e^+e^- scattering within the proto-neutron star, a roughly equal number of all six neutrino types should have been generated. If the tau neutrino, for example, is composed primarily of a massive component then this "neutrino laboratory", in the form of a type-II supernova, provides a copious number of more than 6×10^{57} ν_τ to study (with an equal number of $\bar{\nu}_\tau$).

IV. Radiative Neutrino Decay Model

The computer model developed is a full 3-D simulation including parent neutrino spectra, relativistic kinematics, and angular dependencies. The model allows the neutrinos to stream from the surface of the proto-neutron star and decay in flight. An expanding spherically-symmetric shell is assumed and relativistic kinematics appropriate to two-body decays in flight are used. Relative time since the SN, resultant gamma ray energies, and angle between the gamma-ray arrival direction and the SN-Earth axis are recorded. In this way, arrival-time distributions can be made on selected gamma-ray energy slices for assumed values of m_ν and τ_ν. In addition, energy spectra can be produced on given time windows along with angular distribution histograms.

The parent neutrino spectrum is a Fermi-Dirac distribution with a temperature of 8 MeV [8]. It is higher than the measured $\bar{\nu}_e$ temperature (4 MeV) due to the lower opacity of the proto-neutron star to ν_μ and ν_τ as compared to ν_e. The spectrum is normalized such that the total energy in ν_τ is equal to the measured neutrino energy released from SN1987a in $\bar{\nu}_e$. This is expected to be true due to equipartition arguments.

V. Search Method

Due to the parent neutrino energy spectrum and the assumed finite neutrino lifetime, the energy spectra of the decay gamma rays will evolve over time in a complex (but predictable) manner, examples of which are shown in Figure 1 for two mass values. Because of this spectral evolution, the gamma ray fluence detected - and therefore the mass/lifetime limits obtained - depends on when, in the history of the supernova, an observation occurs. The most sensitive limits can be

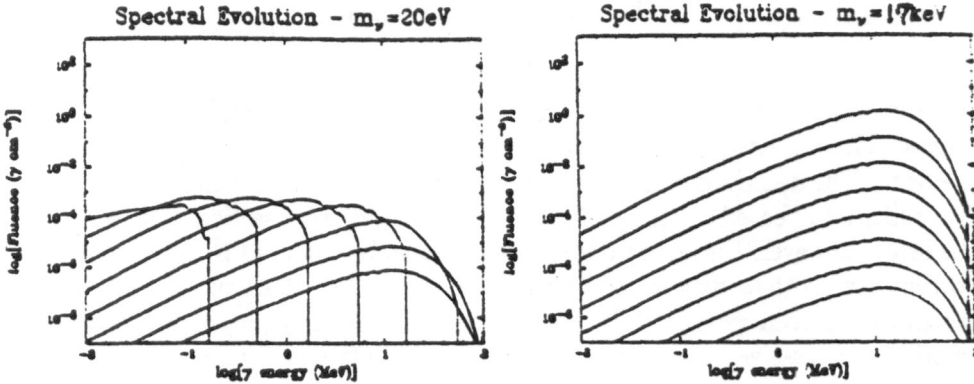

Figure 1: Spectral Evolution - $m_\nu = 20eV$ and $17keV$, D=20 Mpc, $\tau_\nu = 3.5 \times 10^{14}$ (17keV), $\tau_\nu = 5.6 \times 10^{16}$ (20eV), each curve represents the spectrum for a different time interval. The bottom curve represents the spectrum for arrival times 0-10 seconds. The second curve for 10-100 seconds, etc.

achieved by observing the source during a period when the gamma ray fluence is at maximum. The gamma ray arrival times can be distributed over a few seconds or many years, depending on the mass and lifetime of the decaying neutrinos. The arrival time distribution for a sample source is shown in Figure 2.

It is clear that there exists an optimal observational delay which will provide the most stringent limit. In the current search however, the ability to choose the optimal delay does not exist. Except for SN1987a observations which are an integral part of the Compton observatory's viewing program, we rely on past supernovae - or if we're lucky a new supernova - appearing in the COMPTEL instrument's field of view (approximately 64°). Thus, the observational delay since a supernova occurred will dictate the m_ν/τ_ν parameter space being sampled in any given observation. Table 1 shows estimates of the gamma ray fluence (ϕ_γ (cm^{-2})) expected in COMPTEL for different observational delays (assuming an exposure of 5×10^5 s).

VI. Past Searches

No systematic search of this type has been attempted before. However, SN1987A provided a unique opportunity to study neutrino properties. At the time of the supernova there were a number of spacecraft capable of performing gamma ray observations. The most sensitive instrument was the Gamma Ray Spectrometer (GRS) aboard the Solar Maximum Mission (SMM) satellite [9]. No gamma ray pulse was seen within 10 seconds of the IMB neutrino burst time. Thus, limits

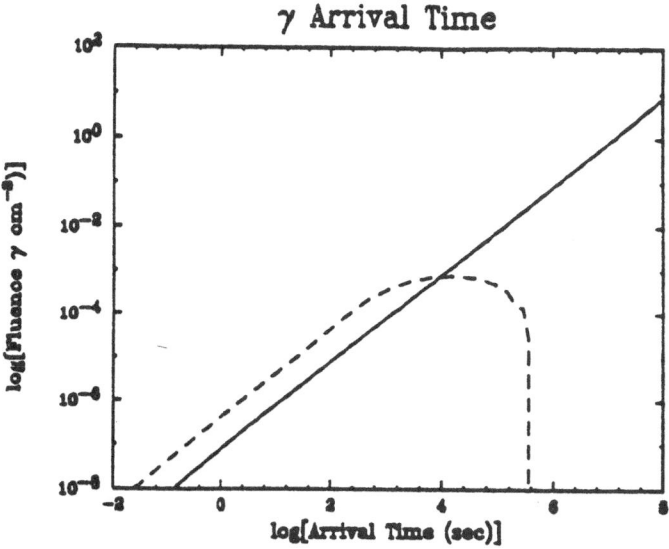

Figure 2: Arrival time distribution of gamma-rays, D=20 Mpc, $\tau_\nu = 3.5 \times 10^{14}$ (17keV,solid), $\tau_\nu = 5.6 \times 10^{16}$ (20eV,dash)

Table 1: Estimates of Gamma-Ray Flux

m_ν	D	0 s delay	10^5 s delay	10^7 s delay
20 eV	55 Kpc	5.6	0.0	0.0
17 keV	55 Kpc	2.4×10^4	2.6×10^4	3.5×10^4
20 eV	20 Mpc	1.6×10^{-2}	1.3×10^{-3}	0.0
17 KeV	20 Mpc	0.17	0.18	0.92

were set on the neutrino lifetime.

The limits published from the SMM observations are based on a simple 1-D model of radiative neutrino decay using monoenergetic supernova neutrinos. For a light neutrino species the limit from Bludman [10] is given by:

$$\tau_\nu = 2.8 \times 10^{15} m_\nu sec \tag{2}$$

while for a heavy neutrino the limit is:

$$\tau_\nu = 6.0 \times 10^{18} m_\nu^{-1} sec \tag{3}$$

both assuming a branching ratio to a radiative decay mode of 1.

As suggested above, the parameter space sampled in such a search is very sensitive to the observational delay and source exposure. Since the supernova occurred. Since the observation window of SN1987a by SMM was short (only 10 seconds due to calibration runs, the SAA, and background considerations), and the observational delay was zero, the sensitivity for observing gamma rays for very massive neutrinos is reduced.

VII. Neutrino Decay and the Diffuse Gamma Ray Emission

Evidence for a diffuse gamma ray flux comes primarily from SAS 2 data [11]. There appear to be multiple components to the measured flux: a galactic and isotropic components. The galactic component is thought to originate from cosmic ray electron brehmstrahlung or Compton radiation, while the origin of the isotropic component is still a significant open question. Many ideas for the production of this isotropic component have been put forward, although none are completely satisfactory [12]. We are currently investigating the role radiative neutrino decay would play in the production of the isotropic diffuse emission.

Assuming an isotropic and homogeneous universe, the energy spectrum of gamma ray emission (due to radiative neutrino decay) from all type-II supernovae since the onset of galaxies can be computed. The shape of the spectrum is dependent on m_ν and τ_ν, as well as the estimated supernova rate per galaxy type, and the density of galaxies in the Universe. By comparing spectral shapes limits can be placed on m_ν and τ_ν, while absolute normalization of the spectrum will place limits on the radiative decay branching ratio. The results of this investigation are forthcoming.

VIII. Summary

This search and analysis is extremely timely due to the recent developments in particle and astrophysics. No systematic study of this type has been attempted before; a non-observation of the gamma ray products from neutrino decay would place significant limits on the mass and lifetime of neutrinos while observation of this process would be a scientific triumph with profound implications for the Standard Model of particle physics, many astrophysical processes, and cosmology.

The time difference between a supernova and the beginning of its observation by COMPTEL will determine the mass/lifetime parameter space sampled. In addition, the source exposure, energy spectra of detected gamma rays, and arrival times will permit analysis of COMPTEL data to look for neutrino decay properties with a sensitivity unobtainable in terrestrial laboratories.

References

1. H.A.Bethe, *Phys. Rev. Lett.* **63** (1989) 837; M.Cherry, *Nature* **347** (1990) 708.

2. J.Bahcall, Neutrino Astrophysics, Cambridge Press, Cambridge (1990).

3. A.Hime and N.A.Jelly, Oxford preprint (1991); B.Sur, *et al.*, LBL Report (1991); J.J.Simpson, *Phys. Rev. Lett.* **54** (1985) 1891; J.J.Simpson and A.Hime, *Phys. Rev.* **D39** (1989) 1825; J.J.Simpson and A.Hime, *Phys. Rev.* **D39** (1989) 1837.

4. L.A.Ahrens, *et al.*, *Phys. Rev.* **D31** (1985) 2732; M.J.Dugan, A.V.Monohar, and A.Nelson, *Phys. Rev. Lett.* **55** (1985) 170.

5. E.Kolb and M.Turner, The Early Universe, Addison-Wesley, New York, (1990) 139.

6. W.Neil Johnson *ed.*, Proceedings of the Gamma Ray Observatory Science Workshop (1989).

7. R.Bionta, *et al.*, *Phys. Rev. Lett.* **58** (1987) 1494; K.Hirata *et al.*, *Phys. Rev. Lett.* **58** (1987) 1490.

8. A.Burrows, Univ. of Arizona preprint no. 90-09; D.Kielczewska, *Phys. Rev.* **D41** (1990) 2967.

9. E.Chupp, W.T.Vestrand, and C.Reppin, *Phys. Rev. Lett.* **62** (1989) 505.

10. S.A.Bludman, to be published *Phys. Rev. D* (1992).

11. C.E.Fichtel, G.A.Simpson,D.J.Thompson, *Ap.J.* **222** (1979) 833.

12. P.V.Ramana Murthy and A.W.Wolfendale, Gamma-ray Astronomy, Cambride Press, Cambridge (1986).

RECENT PROGRESS IN UNDERSTANDING OF X-RAY PULSARS

D.A. LEAHY
Department of Physics and Astronomy,
University of Calgary
Calgary, Canada, T2N 1N4

ABSTRACT. It was soon recognized after the discovery of X-ray pulsars that these systems were neutron stars accreting matter from their stellar companions, either from a stellar wind or via the Roche-lobe overflow mechanism. However detailed understanding of the various aspects of the accretion flow and X-ray emission mechanism has been much slower. Here some areas of progress in observations and in understanding will be discussed. Her X-1 and GX301-2 will be used as examples. Her X-1 is a well-known bright X-ray binary pulsar, and has a 1.70-day orbital period, a rotation period of 1.24 second, and a 35 day semi-periodic cycle in flux. It is in the class of systems accreting via Roche-lobe overflow. GX301-2 is a system with accretion from the stellar wind of the B supergiant companion.

1. Introduction

Binary x-ray sources derive their energy from gravitational potential energy released when matter is accreted onto a compact object. The energy released when a proton is lowered onto the surface of a neutron star, of mass 1.4 solar mass and radius 10 km, is 166 MeV. Thus relatively modest rates of mass transfer onto a compact object can generate large X-ray luminosities. A companion star is generally necessary to provide the matter. For any binary stellar system, the gravitational potential in a coordinate system rotating at the same rate as the axis joining the two stellar centres is called the Roche potential. The equipotential surface which passes through the inner Lagrangian point is called the critical Roche surface. The shape of the critical Roche surface depends only on the mass ratio of the companion star to the compact object: generally it consists of two teardrop shaped surfaces (Roche lobes) whose points touch at the inner Lagrangian point.

Matter can be transferred from the companion to the compact object in one of two ways. It can be given enough kinetic energy so its total energy is above the value of potential energy of the critical Roche surface, or the companion star can expand enough that its surface exceeds the critical surface. The former can be achieved if the companion has a stellar wind. The latter can occur due to expansion caused by stellar evolution. The primary factor determining the volume of the critical Roche surface is the binary period; the mass ratio affects it to a much smaller extent. In either case, an accretion disk is likely to be formed. Observed orbital periods of X-ray binaries range from the ultra-short (11 minutes) to long (of order 1 year). Thus companions that fit within their critical

119

M. M. Shapiro et al. (eds.), Particle Astrophysics and Cosmology, 119–124.

Roche lobes range from sub-dwarfs of about 0.2 solar mass, to supergiants of mass about 60 solar masses.

Here, the subset of X-ray binaries which exhibit X-ray pulsations are considered. The pulsations come from an accreting neutron star which is strongly magnetized and rotating. Thus pulsations are seen due to a local region on the surface, where the accreting matter releases its energy, which rotates in and out of the field of view. For a general review of binary X-ray sources see Lamb (1989), Joss and Rappaport (1984) and references therein. For a review of accretion disks see Pringle (1981). Here, some recent aspects of observations and theory for binary X-ray pulsars are discussed. Recent observations of X-ray pulsars are reviewed by Nagase (1989) and many recent results appear in Frontiers of X-ray Astronomy (1992). Two systems, GX301-2 and Her X-1, are used as examples where the data have allowed new physical phenomena to be studied. GX301-2 is one of the most massive X-ray binary systems known. It has a B2 supergiant companion of 40 solar masses or slightly greater, with a neutron star in an eccentric orbit accreting from the strong stellar wind of the supergiant. In contrast Her X-1 is one of the lower mass X-ray binary pulsars, with a 2 solar mass companion which fills its critical Roche lobe, and feeds mass onto the neutron star.

2. X-Ray Production

The binary X-ray pulsars exhibit a wide variety of pulse shapes, from simple nearly sinusoidal profiles to complex multi-peaked profiles (e.g see White, Swank and Holt, 1983). For some pulsars the pulse shapes are strongly dependent on energy, for others the shapes are nearly independent of energy. There are many factors which govern the observed pulse shape: the geometry of the emission region; the viewing angle to the observer; and the magnetic field strength are three.

In an X-ray binary, material which has crossed the inner Lagrange point into the Roche lobe of the compact object has angular momentum and settles into an accretion disk where it loses angular momentum gradually as it works its way closer to the compact object. For companion stars overflowing the critical Roche lobe, the disk should be large and steady, but for the case of stellar wind fed accretion the disk is expected to be transient and reversing in direction (e.g. Fryxell and Taam 1989). The magnetic fields of neutron stars which are X-ray pulsars are in the range of 1-4 x 10^{12} Gauss (Makashima, 1990) as deduced from their X-ray spectra. Thus at some distance from the neutron star, the Alfven radius, the magnetic pressure will equal the ram pressure of the accreting matter. Outside the Alfven radius, the material pressure dominates, so the accretion disk is intact, but inside the magnetic field dominates (this region is the magnetosphere) forcing the matter to flow along magnetic field lines toward the neutron star (see e.g., Ghosh and Lamb 1979).

The matter is in near free-fall, until near the neutron star surface where it is halted by a shock, after which it enters a post-shock settling zone to the surface. A self-consistent model for this flow including effects of the radiation field has not been found yet. The physics of the production of X-rays within the hot post-shock region on the surface of the neutron star is complex, but with reasonable assumptions can be modelled, and resulting spectra and angular distributions of radiation can be calculated (e.g. Meszaros and Nagel 1985). One must include the effects of the radiation transfer in the strong magnetic field, in particular the cyclotron opacity. With the X-ray luminosity determined by the accretion rate, the Thompson optical depth in this region is given by:

$$\tau_T = 15(L_x/10^{37}erg/s)(\theta_c/0.2)^2 \tag{1}$$

where θ_c is the angular size of the accretion area. The emission region is in the form of a polar cap since the magnetic field lines which connect to the accretion disk beyond the Alfven radius are those nearer the magnetic field axis.

Pulsar X-ray energy spectra have in the past been represented by a power law with exponential cutoff:

$$
\begin{aligned}
F(E) &= A\ E^{-\epsilon} & E < E_c \\
&A\ E^{-\epsilon}\ \exp(-(E-E_c)/E_f) & E > E_c
\end{aligned} \tag{2}
$$

More recently, a formula which includes the effect of cyclotron absorption has been used (in the case given here, up to second harmonic):

$$
\begin{aligned}
F(E) &= A\ E^{-\epsilon}\ \exp(-H(E)) \\
H(E) &= \frac{A_1(WE/E_a)^2}{(E-E_a)^2+W^2} + \frac{A_2(2WE/2E_a)^2}{(E-2E_a)^2+(2W)^2}
\end{aligned} \tag{3}
$$

That this is a good representation of pulsar spectra has been verified for several X-ray pulsars, and thus verifies the strong magnetic field strengths at their surfaces. For example, Mihara et al. (1990) demonstrated convincingly the cyclotron line features in the X-ray spectrum of Her X-1.

Various emission models have been constructed for X-ray pulsars. A geometrical model for pulse shapes (Leahy 1991a) models the emission region on the neutron star by two ring shaped emission regions. Least squares fits were performed to determine the parameters of the emitting rings. It was found that more than half of the 19 pulsars had hollow rings rather than filled caps, and more than half of the 19 had the emission regions offset from being on a single axis, i.e. the magnetic field is not dipolar. Also the mean offset between rotation and magnetic axes was found to be about 0.4 radian. Dermer and Sturner (1991) propose an interesting effect, whereby matter is levitated by radiation pressure above a small region over the polar cap and effectively blocks the emission from the central part of the polar cap. This has the effect of making a polar cap look like a ring.

However in the above works, the effect of gravitational light bending was not incorporated. This has significant effects as shown by Riffert and Meszaros (1988). They calculated beam shapes and pulse profiles for hot spots and for columns on the surface of neutron stars, and find that for small neutron star radii the column produces a pencil beam centred on the opposite pole of the neutron star instead of a fan beam which would be produced in the absence of gravitational light bending. These calculations have been extended to include the effects of scattering of the pulsar radiation by the matter infalling along the accretion column above the polar cap (Brainerd and Meszaros 1991).

3. Reprocessing of X-rays

A luminous X-ray source will illuminate the stellar companion and (if present) its stellar wind, the accretion disk, and any other significant matter in the system (e.g. the accretion stream from the inner Lagrangian point to the rim of the accretion disk). The important processes for X-ray radiation transport through matter are: photoelectric absorption, Compton scattering, atomic K-shell fluorescence (dominated by Fe), and optical emission. For example, optical pulsations were studied to determine that HZ Her (the optical companion to Her X-1) fills its Roche lobe, and thus to determine the mass of the neutron star (Middleditch and Nelson 1976). It was also realized that the reprocessing of X-rays (Lightman and White, 1988, for the case of a planar distribution of matter) would have observable effects on the X-ray continuum spectrum, particularly at energies above about 30 keV where the energy dependence of Compton scattering is no longer negligible.

Of particular interest is the fluorescent iron line emission. The energy of the line and of the iron K-edge absorption (both of these are readily observable in X-ray spectra) depend on the ionization state of iron. This depends on the ionization parameter (L_x/nr^2), thus providing a strong constraint on the density, n, and distance, r, of the matter from the X-ray source. The iron line energy is also affected by Compton, gravitational and doppler shifts. To use this information a good measurement of line shape is needed (not yet achieved) and also modelling of the mass flows in the system needs to be carried out (e.g. for accretion disk profiles see Laor, 1991). Calculation of the line shape due to scattering off a static planar distribution of matter have been done (e.g. George and Fabian 1991).

GX301-2 is an example of a system where reprocessing is important. The supergiant companion is massive (>40 solar masses) and has a strong stellar wind. The X-ray spectrum has a continuum of the form given above for accreting strongly-magnetic neutron stars, with the additional features of strong iron line emission and strong iron edge absorption (Leahy et al. 1989a,b). The iron line is of fluorescent origin and follows the theoretical relation between equivalent width and column density for spherically distributed matter for this system, but with additional scatter. The iron line intensity - continuum intensity relation is not linear, and the iron line energy - column density decreases. These indicate a changing ionization state of the fluorescing material with amount of material around the source and with the intensity of the source.

The shape of the pulse from the neutron star also gives information on nearby matter. If the matter is extended more than cP, with P the pulse period, the scattered radiation will be constant with pulse phase. Thus a decrease in pulse modulation can be related to the amount of scattered radiation. For GX301-2 the conclusion is that modulations are generally consistent with the observed column densities, but that the pulse shape changes must also have a contribution from the intrinsic variations in pulse shape (this conclusion is independent of the angular distribution of radiation from the neutron star).

The neutron star serves also as a probe of the stellar wind. GX301-2 has an eccentricity of 0.47 and shows changes in luminosity and column density with orbital phase caused by the varying distance to the B supergiant companion. This cycle in brightness has been modeled (Leahy, 1991b) using a stellar wind plus gas stream model. Modelling using a physical wind model, in which a gas stream is caused by the gravitational wind enhancement by the neutron star in an eccentric orbit (Stevens, 1988) is in progress.

4. Her X-1, A Prototype X-Ray Binary

Her X-1 is one of the most studied X-ray pulsars, with the companion HZ Her filling its Roche-lobe. For example, studies of the X-ray spectral and intensity changes during ingress and egress of the eclipses show that the atmospheric scale height of HZ Her is variable (Day et al. 1988). The 35 day cycle in intensity, with a main-on phase and a short-on phase separated by low states has been well documented. The main model for this cycle is a precession of HZ Her, which results in periodic mass transfer and in a tilted precessing accretion disk (Petterson, 1977). Evidence for the periodic mass transfer and precession of HZ Her came from analysis of timing of the pre-eclipse dips (Crosa and Boynton, 1980).

Extensive observations of Her X-1 were carried out with the GINGA LAC (Large Area proportional Counters) (Makino et al. 1987). The large collecting area results in much better spectra and time resolution from Her X-1 than previously possible. The pre-eclipse dips have been seen with 4 s time resolution, and show fast (approx 20 s) factor 2 excursions in intensity which occur throughout the dip period. Thus whatever causes the dips must be small. A dip period covering 3 hours has been shown to exhibit 144 s periodic flux changes (Leahy, 1992). The spectrum of Her X-1, both during dips and throughout the 35 day cycle shows two continuum components, an unabsorbed component and a highly absorbed (of order 10^{24} cm^{-2} column density) component (Leahy et al. 1991). This may be interpreted in terms of a partial covering model or in terms of a primary source and electron-scattered X-rays from a hot corona. Existing GINGA data on the pulse shapes, their dependence on energy and the changes with time for Her X-1 are sure to challenge theorists and modellers.

5. Summary

Binary X-ray pulsars are being studied extensively using X-ray timing and spectral data. As a consequence, more realistic physical models of processes occuring in these systems can be constructed and tested. So our understanding of X-ray pulsars is improving steadily. In the near term, new instrumentation with higher spectral resolution and better sensitivity will allow new studies to be done. In particular, measurements will be made of the shape and variability of the fluorescent K-shell lines, and significant progress on the properties of circumstellar material and accretion flow is expected.

References

Brainerd, J. & Meszaros, P. 1991, ApJ, 369, 179
Crosa, L. & Boynton, P. 1980, ApJ, 235, 999
Day, C., Tennant, A. & Fabian, A. 1988, MNRAS, 231, 69
Dermer, C. & Sturner, S. 1991, ApJ, 382, L23
Frontiers of X-ray Astronomy, ed. Y. Tanaka & K. Koyama, 1992 (Tokyo: Universal Academy Press)
Fryxell, B. & Taam, R. 1989, ApJ, 335, 862
George, I. & Fabian, A. 1991, MNRAS, 249, 352
Ghosh, P. & Lamb, F. 1979, ApJ, 234, 296
Joss, P. & Rappaport, S. 1984, ARAA, 22, 537

Lamb, F. 1989, in 'Timing Neutron Stars', ed. H. Ogelman & E. van den Heuvel (Kluwer Academic Publishers), p.649

Laor, A. 1991 in Iron Line Diagnostics in X-ray Sources, ed. A. Treves, G. Perola, L. Stella (New York: Springer-Verlag) p205

Leahy, D. 1991a, MNRAS, 251, 203

Leahy, D. 1991b, MNRAS, 250, 310

Leahy, D. et al. 1992, in The Compton Observatory Science Workshop (NASA Conference Publication 3137) p193

Leahy, D. et al. 1991, 22nd International Cosmic Ray Conference (Dublin Institute for Advanced Studies) p13

Leahy, D. Matsuoka,M. Kawai,N. & Makino,F. 1989a, MNRAS, 236, 603

Leahy, D. Matsuoka,M. Kawai,N. & Makino,F. 1989b, MNRAS, 237, 269

Lightman, A. & White, T. 1988, ApJ, 335, 57

Makashima, K. 1990 in Proceedings of Structure and Evolution of Neutron Stars, Nov. 6-10, 1990, Kyoto.

Makino, F. et al. 1987, Astrophys.Lett.Comm., 25, 233

Meszaros, P. & Nagel, W. 1985, ApJ, 299 138

Middleditch, J. & Nelson, J. 1976, ApJ, 208, 567

Mihara, T. et al. 1990, Nature, 346, 250.

Nagase, F. 1989 in Proc. 23rd ESLAB Symposium (ESA SP-296, Nov.1989) p45

Nagase, F. 1991 in Iron Line Diagnostics in X-ray Sources, ed. A. Treves, G. Perola, L. Stella (New York: Springer-Verlag) p111

Parkes,G. et al 1980, MNRAS, 191, 547

Pringle, J. 1981, ARAA, 19, 137.

Riffert, H. & Meszaros,P. 1988, ApJ, 325, 207

Shapiro, S. & Teukolsky, S. 1983, Black Holes, White Dwarfs and Neutron Stars (New York: Wiley)

Stevens, I. 1988, MNRAS, 232, 199

White, N., Swank, J. & Holt, S. 1983, ApJ, 270, 711

Simultaneous Air Shower and Deep Muon Observations at Soudan

W. W. M. Allison[3], G. J. Alner[4], I. Ambats[1], D. S. Ayres[1], L. Balka[1], G. D. Barr[3], W. L. Barrett[1], D. Benjamin[5], P. Border[2], C. B. Brooks[3], J. H. Cobb[3], D. J. A. Cockerill[4], K. Coover[1], H. Courant[2], B. Dahlin[2], U. DasGupta[2], J. W. Dawson[1], V. W. Edwards[4], D, Demuth[2], B. Ewen[5], T. H. Fields[1], C. Garcia-Garcia[4], R. H. Giles[3], M. C. Goodman[1], R. Gray[2,] S. Heppelmann[2], N. Hill[1], J. H. Hoftiezer[1], D. J. Jankowski[1], K. Johns[2], T. Joyce[2], T. Kafka[5], J. A. Kochocki[5], P. J. Litchfield[4], N. P. Longley[6], F. V. Lopez[1], M. Lowe[2], W. A. Mann[5], M. L. Marshak[2], E. N. May[1], L. McMaster[5], R. Milburn[5], W. Miller[2], A. Napier[5], W. P. Oliver[5], G.F. Pearce[4], D. H. Perkins[3], E. A. Peterson[2], L. E. Price[1], D. Roback[2], D. B. Rosen[2], K. Ruddick[2], B. Saitta[5], J. L. Schlereth[1], D. Schmid[2], J. Schneps[5], S. Schubert[2], P. D. Shield[3], M. Shupe[2], N. Sundaralingam[5], M. A. Thomson[3], J. Thron[1], L. M. Tupper[3], S. Werkema[2], N. West[3], C. A. Woods[4]

[1] High Energy Physics Division, Argonne National Laboratory, Argonne, IL 60439, U.S.A.

[2] School of Physics and Astronomy, University of Minnesota, Minneapolis, MN 55455, U.S.A.

[3] Department of Nuclear Physics, University of Oxford, Oxford, U. K. OX1 3RH

[4] Rutherford Appleton Laboratory, Chilton, Didcot, U. K. OX11 0QX

[5] Physics Department, Tufts University, Medford, MA 02155, U.S.A.

[6] Department of Physics and Astronomy, Carleton College, Northfield, MN 55057, U.S.A.

Recently, a few groups[1] have made attempts to simultaneously measure the electromagnetic and deep underground muon components of cosmic ray showers. This is one such experiment, using a small proportional tube surface air shower detector in coincidence with the Soudan 2 nucleon decay detector. It has run since June, 1991, and some 22 000 events which triggered both detectors have been recorded. This body of data may provide insight into the composition of cosmic rays in the "knee" area of the cosmic ray spectrum.

I. Experimental Apparatus

The Soudan 2 proton decay detector[2] is a large tracking drift calorimeter located 715 m (2090 mwe) beneath the surface in a non-operational iron mine in Soudan, Minnesota, U.S.A. (latitude 47.82° N, 92.25° W). It is still under construction, but during this first year of surface/underground operation, it was approximately 10m x 8m x 5m high, with a mass of some 720 tons.

The Soudan 2 surface detector is a small (40 m^2) proportional tube array on the surface 50 m south and 140 m east of the underground calorimeter, corresponding to a zenith angle of approximately 12°. It is made of thirty-two (32) 20cm x 5cm x 6.7m long proportional wire modules, arranged in two groups of 16 each. Each individual module contains 15 separate 2.5 cm square proportional tubes arranged in two parallel overlapping layers. The 480 individual wires are ganged into 240 channels and operated in an on/off mode. They have no spatial resolution in their length.

125

M. M. Shapiro et al. (eds.), Particle Astrophysics and Cosmology, 125–127.
© 1993 Kluwer Academic Publishers.

This arrangement provides a measure of the average particle density at the surface detector for each event that triggers Soudan 2. The underground muon track defines the direction of the cosmic ray shower. The perpendicular distance from the shower core to the surface detector and density of particles allow one to approximately determine the surface shower size through well-known cosmic ray shower profiles[3].

II. Data

For each cosmic ray muon event which triggers the Soudan 2 detector, a 128 µs long window is examined in the surface detector data. Events with two or more separate two-layer tracks are retained. The time for an underground muon to have passed nearest the surface detector is calculated, and the relative time distribution of events is shown in figure 1. There are 25 944±161 events in a 3 us window around the expected coincident time, above an out-of-time background of 3 746±61 per 3 us. The signal width is consistent with the ±1 µs timing width of the surface detector.

Figure 2 is a contour plot of the cosmic ray muon/surface intersection point for 19 745 events within the 3 µs coincident window. The origin is directly above Soudan 2, and north is roughly along the +z axis. Note that the coincident events tend to cluster strongly about the surveyed position of the surface detector. Figure 3 contains 16 142 out-of-time events. They have no such tendency. The in-time, length-selected events (2 m<track length<8 m) in this plot are approximately three-quarters single muons and one-quarter multiple muons.

The Soudan 2 surface detector first began recording data in May, 1991, and has been fully operational since mid June, 1991. It has run regularly (except for Soudan 2 construction) since that date, during which time it has collected coincident above- and below-ground cosmic ray events at the rate of approximately 50 000 per live-year (6 per hour). The data at present (as of 29 May, 1992) number some 22 000 events, of which just over 5 000 exhibit multiple underground muon tracks. The large and increasing size of this data set may allow it to be useful in efforts to determine the composition of ultra-high energy cosmic ray primaries.

[1]U. Das Gupta, T. H. Fields, and K. Ruddick, *Composition Studies Based on Coincident Air Shower Array and Underground Muon Data*, Proceedings of the 22nd International Cosmic Ray Conference, OG 6.1.14 (1991).

EAS-TOP and MACRO collaborations, *Study of High Energy Cosmic Rays Through the Measurement of the Electromagnetic and TeV Muon Components of Extensive Air Showers by EAS-TOP and MACRO*, Proceedings of the 22nd International Cosmic Ray Conference, OG 6.1.23 (1991).

W. W. M. Allison, *et al.*, *Simultaneous Observations of Extensive Air Showers and Underground Muons at Soudan 2*, Proceedings of the 22nd International Cosmic Ray Conference, OG 6.1.24 (1991). The above is essentially an update on the Soudan 2 surface/underground initiative introduced at the 1991 conference in Dublin.

[2]J. L. Thron, *et al.*, *The Soudan 2 Proton Decay Experiment*, Vienna Wire Chamber Conference, NIM 283, 642 (1989).

[3]K. Griessen, Ann. Rev. of Nuc. Sc., 10, 63 (1960).

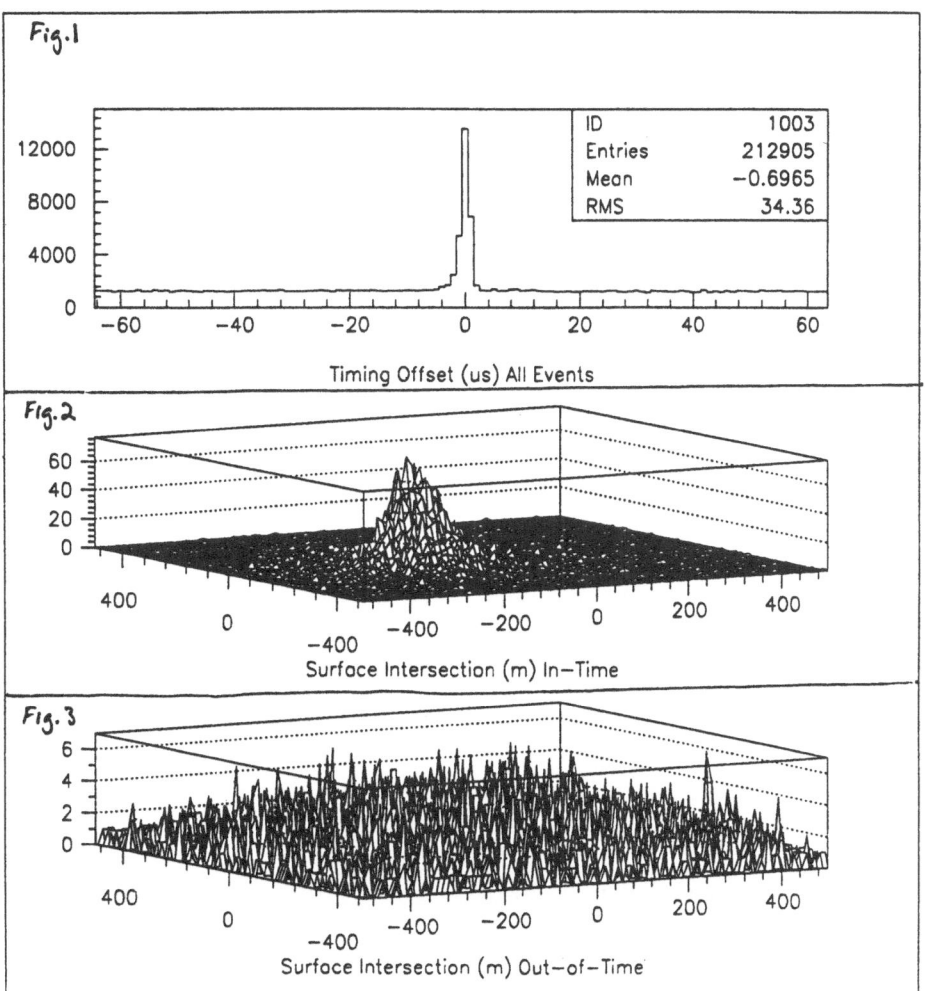

Fig.1

ID	1003
Entries	212905
Mean	−0.6965
RMS	34.36

Timing Offset (us) All Events

Fig.2

Surface Intersection (m) In−Time

Fig.3

Surface Intersection (m) Out−of−Time

NUCLEOSYNTHESIS IN THE EARLY UNIVERSE

W.A. Fowler
California Institute of Technology
Kellogg Lab. 106-38
Pasadena, California 91125

ABSTRACT. Calculations on early nucleosynthesis in the inhomogeneous, inflationary universe yield the universal abundance of the elements from hydrogen to lithium in agreement with observation except that too much helium is produced. In this paper this problem is discussed in some detail. In addition, nucleosynthesis of the heavy elements takes place in the neutron rich sea of the inhomogeneous universe. Rapid neutron capture and fission cycling are found to produce the low abundances of the heavy elements found in the oldest stars in the Galaxy. Numerous problems are pointed out which involve the amount of luminous matter in the early universe.

This lecture is a modified version of a lecture given at Nobel Conference XXVII, Gustavus Adolphus College, St. Peter, MN., October 2, 1991. In my Nobel lecture in 1983 I addressed the question of the quest for the origin of the chemical elements in our observable universe; quite simply it was argued that isotopes of hydrogen, helium and lithium were produced from protons and neutrons in a homogeneous and isotropic early universe commonly called "The Big Bang." Still heavier elements and their isotopes were synthesized from the primordial hydrogen and helium by nuclear processes in stars, novae and supernovae. It is still believed that the great majority of the elements beyond helium are produced in these astrophysical circumstances. However, since the suggestion by Alan Guth in 1981 that our early universe experienced an enormous growth in size or *inflation*, it has come to be believed that a small fraction of the heavy elements were produced shortly after this inflation. This explained among other things why the oldest stars in the Galaxy observed by astronomers show evidence for small but significant amounts of heavy elements in their spectra which could not have been made in stars because there were not any stars previous to these oldest ones.

Studies of this new development were greatly enhanced when Edward Whitten in 1984 showed that the universe after inflation would be inhomogeneous rather than homogeneous. Our observable universe, expanding as Edwin Hubble found from his red shift measurements in the 1920's, could be thought of as an expanding bubble of matter as we know it, into an otherwise steady state universe consisting of extremely high density stuff called vacuum matter, for want of a better term. This vacuum matter corresponded to Einstein's cosmological constant and to Friedmann's cosmological term in his equations governing the behavior of our observable universe with a finite but rapidly increasing radius. Remarkably, this vacuum matter exerted negative pressure and thus

129

M. M. Shapiro et al. (eds.), Particle Astrophysics and Cosmology, 129–134.
© 1993 *Kluwer Academic Publishers.*

130

can be thought of as the cause of our expanding bubble.

The matter in our expanding bubble was inhomogeneous since the early quark-gluon plasma transformed into a hadron gas consisting ultimately of high density proton rich regions immersed in a neutron rich sea. Nucleosynthesis in the neutron rich sea permitted highly charged, heavy nuclei to be synthesized since neutrons are neutral with zero electric charge. Coulomb repulsion prevents charged protons and alpha particles from amalgamating with highly charged nuclei. There is no Coulomb repulsion for neutrons. Many authors, including myself and my collaborators, have contributed to what is by now a fairly clear picture of nucleosynthesis in an inhomogeneous universe. Experimentalists have studied in the laboratory some of the many additional reactions which took place in the neutron rich sea which was also rich in deuterium, tritium and helium nuclei. This past decade has been a very exciting one in contributing to our knowledge of the early history of the universe we inhabit. There is still more to be done and that is mainly what is stressed in what follows.

The inflationary universe is the background for what follows. For those who are interested in the origin of inflation it will be found in Guth's article in Phys. Rev. D23, 347 (1981), and in A. Linde, Phys. Lett. 108B, 389 (1982). A major point is that the early universe before inflation was very small with all parts in causal contact and thus at the same temperature. That led to the fact that the cosmic background radiation is at the same temperature everywhere today, as observed, to better than one part in ten to the fifth. Previously in our early ideas of the big bang, we really had no way of understanding the constancy of the cosmic background radiation but when you start out with something small enough which was all in causal contact, thermal equilibrium was established and continued during the expansion. The early exponential expansion during 10 to the minus 32 seconds, led to a flat euclidean universe with the curvature parameter, which appears in Friedmann's equation going to zero. If one expands a sphere with lines of longitude drawn on it and looks at a very small part of the sphere, the lines of longitude become parallel and the curvature which is described by the constant k in the Friedmann equations goes to zero. This is one of the basic things which comes out of inflation. I've always liked this because it makes Friedmann's equations much easier to solve than when k is positive or negative. Thus there will be a universal expansion rate and I will take the capital letter A as a distance scale factor in describing the expansion.

Hubble's constant is equal to the time derivative of A $(=\dot{A})$ divided by A, that is $(\dot{A}/A)$. Friedmann's equation yields values for $8\pi G\rho_t/3$ where G is the gravitational constant, and ρ_t is the mean matter density in the universe. This density, ρ_t, includes the density ρ of real matter and the density ρ_v of vacuum matter, which is assumed to fill all space uniformly. If one sets the curvature parameter equal to zero, as discussed previously in connection with inflation, then one obtains

$$H^2 = (\dot{A}/A)^2 = (8\pi G/3)\rho_t = (8\pi G/3)(\rho + \rho_v)$$

This series of equalities can be simplified by setting

$$\Omega = 8\pi G\rho/3H^2 \quad \text{and} \quad \lambda = 8\pi G\rho_v/3H^2$$

so that

$$\Omega + \lambda = 1,$$

which for

$$\lambda = 0$$

yields

$$\Omega = 1.$$

This Ω can be expressed as

$$\Omega = \Omega_b + \Omega_e$$

where the subscript b designates baryons or ordinary matter while the subscript e designates exotic particles such as neutrinos, photinos, axions, Higgs bosons, etc.etc. The standard Big Bang production of ^2H, ^3H, ^4He, and ^7Li abundances required $\Omega_b \approx 0.1$ as noted by Robert V. Wagoner, William A. Fowler and Fred Hoyle as early as 1967 in Ap.J. **148**, pp.3 to 49. With $\Omega = 1$ these results yield $\Omega_e \approx 0.9$. This large amount of exotic matter has never appealed to me. I have enough exotic things in my life without having so many particles around that are exotic in great numbers. Could $\Omega_b = 1$? My answer and that of other people is *yes* from studies made in the 1980's. For example refer to R.A. Malaney and W.A. Fowler, American Scientist 76,472 (1988).

Now, more about the inflationary universe. The total energy of the universe is equal to the rest-mass energy plus the kinetic energy plus the potential energy. The kinetic energy is just one-half mass times the velocity squared and the potential energy is the gravitational constant times any mass squared over the separation distance. The curvature term in Friedmann's equation creates difficulties in this regard. However, with k equal to zero one finds that the total energy in the universe is zero and it has always pleased me that in this general picture, the conservation of energy is observed. So inflation reduces the curvature parameter to zero and thus the total energy of the observable universe is zero. Energy is conserved in the universe that we inhabit. There is no problem in the conservation of energy in the expansion after inflation. The expansion of the universe, follows the conservation of energy or, in other words, is driven by by the conservation of energy.

It is worthwhile to repeat the consequences of the inflationary model. First of all, the total energy of the universe is equal to zero. The rest mass energy plus the kinetic energy of expansion minus the gravitational potential energy is zero. Then it follows from Friedmann's equations that time in the universe is governed by a relation between the age of the universe and the Hubble time. Hubble's law yields velocity proportional to distance. If one brings the distance over on the side of the equation where the velocity is, velocity divided by the distance is time and that we call the Hubble time. It can be shown that the age of the universe is two-thirds of the Hubble time when the curvature term in Friedmann's equation is zero. I must note that there is a great deal of controversy about this simple "time" relationship. I think it is true and many of my colleagues who had worked on the problem think it is true but I want to emphasize that there are skeptics who would challenge even this simple relation which holds for a zero curvature universe.

Now consider a possible scenario based on the inflationary model which is very oversimplified. We can look back in time as we look out in distance so we can look back to the period when the universe was opaque. It was at the end of that period that the cosmic black body radiation was created. It started out way back then at about 4000° kelvin and now due to expansion and the accompanying cooling it has come down to 2.736° kelvin as is accurately measured. The observable horizon is the velocity of light times the Hubble time but the true horizon can be much larger. In fact, if you wish to dream a bit, you can think that there may exist other universes immersed in the "false" vacuum. The false vacuum is equivalent to a large cosmological constant. It obeys Friedmann's equation. No question about that. It just has a large cosmological constant and what we hope is that here, right around us, the cosmological constant is zero. So that's a possible scenario. By the way, those other universe have to be at great distances from us and each other because in the roughly 11 billion years of the age of our universe, we don't want these other universe overlapping. The consequences would be spectacular and disastrous not only for the overlapping universe but for all other universes including ours.

In the very beginning of our observable universe the temperature was so high that the protons and neutrons were broken down into their constituent particles which are called "quarks." As things cooled a quark-hadron phase transition occurred. Three quarks fused together to make hadrons. There were some heavier hadrons than the baryons which gradually admitted gamma rays or various other things and came down to the baryons which make up the stuff like us. We consist of the lowest mass-energy type of the hadrons. Many authors have shown that the early quark-hadron phase transition makes nucleosynthesis in a closed baryonic universe in fair agreement with observation of the primordial abundances of deuterium 2, helium 3, helium 4, lithium 7, beryllium 9, and a small amount of the elements with atomic number greater than 12. The key is that the baryon density fluctuation leads to a neutron rich region in which part of the Big Bang synthesis takes place. As noted previously there is no Coulomb repulsion between neutrons so in the neutron rich region nuclear reactions can involve more heavily charged nuclei and produce nuclei with atomic numbers greater than 12. In fact, the reactions can proceed all the way up to uranium and thorium even though only a very small number of heavy nuclei is made at this stage. Those who follow my line of thought think that this is where the small amount of heavy elements that astronomers see in the very oldest stars, stars in the oldest galaxies, came from.

I now turn to what Robert A. Malaney, now at CITA, the University of Toronto, did with me looking over his shoulder more or less. We decided to investigate Big Bang nucleosynthesis not only with $\Omega = 1$ but with the Ω being made up of baryons, that is $\Omega_b = 1$. Following the quark-hadron phase transition there existed two distinct types of regions, again an over simplication. There were high density regions which were very proton rich. They were much like the whole universe in the standard homogeneous model. In these regions there are about 6 times as many protons as neutrons, because neutrons are heavier than protons. In the ratio of protons to neutrons, there is an exponential term that contains the mass difference to a negative power so protons turn out to be 6 to 7 times as abundant as neutrons. That is the case in the bubbles where all the matter started. But neutrons diffuse much faster than protons. When a proton is moving it can collide with another charged particle and be scattered usually with some energy loss. Neutrons are not scattered by charges so they "waffle" around on their own. Finally, they escape from the high density bubbles and produce a low-density region which is neutron rich and which surrounds the proton rich bubbles. There is a different nucleosynthesis in the two regions and the result produces the primordial abundances of deuterium 2, helium 3, helium 4, and also lithium 7 in the observed primordial amount if back diffusion of neutrons destroys beryllium 7 in the proton rich regions. This will be discussed in more detail in what follows. The most important result is that a small amount of heavy elements, mostly from A equals 12 up to about 60, is made in what is called the r-process which also occurs in stellar nucleosynthesis. The "r" comes from rapid in the rapid neutron capture process. We now see that r-process nuclei can be produced in the neutron-rich sea. The neutrons will produce nuclear reactions all the way to thorium and uranium and then fission occurs. Out of that fission there emerge two nuclei where before there was only one. So, the whole thing works beautifully in that you get more light nuclei to which you can add neutrons and so forth and so on. These nuclei form the seeds for the s-process in first generation stars. The lower case "s" means the neutrons are captured slowly, much more slowly in stars in the s-process than in the early universe.

We can go on with this. In this over simplified scenario there is a proton rich bubble in a neutron sea. It turns out that the neutrons must not be used up too soon. That is all well and good because the bubbles originally are quite dense and have very little surface so there is very little neutron penetration. Then as the expansion goes on the density in the bubbles gets lower so neutrons are not scattered at the surfaces so much. In the expansion the surface area gets greater so the neutrons in the neutron rich sea can reenter the bubbles. That is all to the good because in these proton rich bubbles one of the nuclei produced is beryllium 7. Beryllium 7 has 4 protons and 3 neutrons. That is the nucleus which occurs at mass 7 in the proton rich region. Four protons compared to 3 neutrons. The stable form of mass 7 is lithium 7 which has 3 protons and 4 neutrons. When Malaney and I first made our calculations we had a "terrible" result because too much beryllium 7 had been produced and finally the beryllium 7 decayed by electron capture to lithium 7 so too much lithium 7 was produced. The solution to this problem came when neutrons diffused back into the bubbles hit the beryllium 7, changed it into lithium 7 plus protons. With the protons in the bubbles the lithium 7 hit them and went to two helium nuclei so the result was not too much beryllium 7 and ultimately not too much lithium 7.

One of the things that pleased me in this work was that the inhomogeneous neutron rich regions required the study of many additional nuclear reactions in the laboratory which we didn't require in the old homogeneous universe. In the inhomogeneous universe, the neutron rich region brought the necessity for studying, in the laboratory, many new reactions not previously required. There are quite a few of these new reactions and that has one been one of the things that pleased me most that some work that we were doing theoretically could lead to work to be done in the laboratory because by this time work on stellar nucleosynthesis in the laboratory had pretty well petered out, So this was a new enterprise for Ralph Kavanagh and Charles Barnes and others in our laboratory and other places.

In the old homogeneous universe one obtained deuterium 2, helium 3, helium 4, and lithium 7 with the observed abundances when the density was about 3×10^{-31} g cm^{-3}, which is much less than the critical density at 6×10^{-30} g cm^{-3}. So, in the old point of view, in a homogeneous universe as many other people have found, the density of the universe as measured from what went on in the early Big Bang or just afterward, was roughly one-tenth of the critical density. Since we had come to believe that there was something that was making up the critical density, we had to do it with all kinds of exotic particles like neutrinos, photinos, axions, Higgs bosons.... I never liked making the critical density of the universe with these exotic particles so that partly motivated me to go to the study of the inhomogeneous universe. In this case one has a lot of flexibility because there exists a proton rich region in which nucleosynthesis can take place and also there exists a neutron rich region in which nucleosynthesis can take place as discussed above.

To make a long story short, one of the nice things about the inhomogeneous model is there are some quantities which can be varied. One is the fraction of the volume in the neutron rich region relative to the fraction of the volume in the proton rich region. Malaney used a computer to search for the best fits to the abundances and found another adjustable variable in the rate of the neutron diffusion back into the proton rich region. It is very difficult to calculate neutron diffusion. Anyone who's in the field knows that, so all we did was say that the ratio of the neutrons in the proton rich region relative to the neutrons in the neutron rich region was the quantity A_0. A large A_0 indicates very rapid back diffusion. A_0 equals zero signifies no back diffusion. The calculations showed that A_0 greater than about 0.3 gives quite good agreement with the observations. The observed deuterium is somewhat greater than 5×10^{-6}. The calculations yield approximately 10^{-5} for $A_0 = 0.3$. The observed helium 3 is less than 3×10^{-4}. The calculations yield 3×10^{-5} which does not contradict the observations. The helium 4 abundance is a problem. The calculations yield too much helium $4 = 0.25$. It depends on how serious you take the people who think they can specify the amount of helium when our galaxy formed. Originally there was some latitude in the observations. Now the abundance has been tied down to 0.228 ± 0.005 by Bernard Pagel and A. Kazlauskas and if that's the case the calculations are high. But these calculations are very oversimplified and I am still working on this problem with others in our laboratory. We're trying to see if there isn't some way to get the calculated number for helium 4 down to a lower value. In the Large Magellanic Cloud the abundance of a lithium 7 runs between 2×10^{-10} and 8×10^{-10}. The calculations for $A_o = 0.13$ fit this range. So, in general this picture of nucleosynthesis in the inhomogeneous universe agrees fairly well with the observations except for the fact that too much helium 4 is made and that's fun because it leaves some more work to be done.

We turn to Big Bang nucleosynthesis of the heavy elements. It all takes place in the neutron rich sea. The nucleosynthesis goes quickly up to ^{18}O with neutrons, deuterons, tritium and alpha particles. In addition rapid neutron capture occurs all the way up to fissionable nuclei. New seed nuclei are produced. Moreover fission produces additional neutrons, about three per fission process. Thus fission cycling can occur. Malaney and I found that 25 cycles would do the trick and that's another arbitrary number but a reasonable one. Fission cycling produces the low abundances of the heavy elements (10^{-4} solar) found in the oldest stars in the galaxy so our galaxy formed with a small amount of heavy elements produced in inhomogeneous Big Bang nucleosynthesis.

Numerous problems remain. The luminous matter consists of baryons. This luminous matter is about one tenth of the critical density and if $\Omega_b = 1$ then what is the dark or missing matter? I believe it is baryons but they have to be dark and that's one of the problems. There's another problem; astrochronological calculations yield Galactic globular cluster ages up to 16 billion years. That is greater than my Galactic radiochronological age of 10 billion years. One way to reduce globular cluster ages is to have main sequence mass loss when stars first form and that has been studied by Lee Ann Wilson at Iowa state. In addition D.E. Winget and his collaborators find that the age of the oldest Galactic white dwarfs is $\lesssim 10$ billion years so the astrochronological globular cluster ages may be high.

The epitome is we may well live in the simplest of all Einstein's universe. His curvature parameter is zero. His cosmological constant is zero. His space time is Euclidean. His universe has zero total energy and his matter is stuff like us. I think Einstein would like that; I know I do and I hope you do. I remain at work in this field driven by this hope.

THE AGE OF THE OBSERVABLE UNIVERSE

W.A. FOWLER
California Institute of Technology
Kellogg Lab. 106-38
Pasadena, CA 91125

ABSTRACT. The inflationary model for the inhomogeneous universe predicts that the age of the universe is exactly two-thirds of the Hubble time, the inverse of the Hubble constant. In this paper the Hubble time is taken to be 17 billion years from red-shift measurements and the age of the observable universe is found to be 11 billion years using radioactive chronometers. Since 11/17=0.647 is in close accord with 2/3=0.667, the inflationary model calculations can be taken as substantially correct in this case.

This lecture is an up-dated version of a lecture given at the Wright Science Colloquia, Geneva, Switzerland on September 17, 1986, which was based on the Tenth Edward Arthur Milne Lecture given at Oxford University, Oxford, England on November 13, 1986, and published in the Q. Jl. R. astr. Soc. 28, 87(1987). Our planet earth is part of what we call the solar system. The sun, the earth, the moon, and all the other planets and satellites in the solar system are 4.6 billion years old give or take one-tenth of a billion years or so. But even that enormous age—incomprehensible to us in terms of human experience—is somewhat less than one half of the age I will be treating in this lecture, the Age of the Observable Universe. You will find that I have a bias in favor of using radioactive nuclei as chronometers to determine cosmological ages just like geologist use them to date rocks, the meteorites and even the moon, and archeologists and anthropologists use them to date their specimens. My conclusion will be that the Universe is approximately 11 billion years old. 11 billion years is 350 million billion seconds. I am reminded of an old Gaelic Proverb I learned from a student companion on a walking trip in the Highlands of Scotland: *When God made time, He made plenty of it.* By the way, by billion I mean one thousand million in spite of my Scot-Irish-English heritage!

In this lecture I am going to talk about two chronological matters. First of all, I will discuss the time scale characteristic of the expansion of the Universe. The expansion of the Universe was established by Edwin Hubble using red-shift and luminosity measurements on the light from distant galaxies. Out of his observations came Hubble's Law and the concept of the Big Bang cosmology in which the expansion of the Universe is said to have started in a Big Bang. I will state Hubble's Law as simply and as naïvely as possible—the velocity of recession of a distant galaxy is proportional to its distance from us and the constant of proportionality is called the Hubble constant. It's reciprocal has the dimensions of time and is called the Hubble time. I believe that

M. M. Shapiro et al. (eds.), Particle Astrophysics and Cosmology, 135–138.
© 1993 *Kluwer Academic Publishers.*

the Hubble time is slightly less than 17 billion years. I realize full well that there are many provisos and qualifications necessary in describing the law correctly in the language of general relativity. At large red-shifts, gravitational red-shifts must be taken into account. On the simple point-of-view the red shift of the light from a Galaxy moving away from us is a Doppler shift to longer wave lengths just like the sound from a source moving away from us has a longer wave length and thus has a deeper pitch. As a steam locomotive buff it intrigues me that the Austrian physicist, Christian Johann Doppler, put a musical band on a moving flat car pulled by a steam locomotive to prove experimentally the shift in pitch which is named for him.

There are several features of the Expanding Universe which are somewhat difficult to comprehend. First of all, there is no center of the expansion. All space is expanding. If we wish to think we are at the center then the inhabitants of a distant Galaxy may also think they are at the center. A simple analogy in two spatial dimensions is the surface of a balloon. Imagine the surface to be dotted with ink spots. As the balloon is blown up all the spots see the other spots receding from them - no spot is the central spot! Any coordinate system drawn on the balloon expands. That is what we mean by all space is expanding. Furthermore the thickness of the balloon material decreases as the balloon is blown up. This is analogous to the fact that the average density of matter is decreasing everywhere in the Universe. I emphasize average since the earth is not expanding, our solar system is not expanding, our Galaxy is not expanding. They are all held together by gravitational forces. The universal expansion is large scale - galaxies are moving apart and the average number in a large, given volume in the Universe is decreasing with earth time.

In the second part of my talk, I will turn to the Age of the Chemical Elements which uses the production and decay of the radioactive elements as time keepers. Remember, the Big Bang produced only a very small fraction of the elements beyond helium in the solar system. The heavy elements were almost entirely produced over the lifetime of our Galaxy, the Milky Way, by stars which were born, aged, and died before the formation of the solar system and in their death throes spewed out heavy elements into the interstellar medium. When the Solar System condensed out of the interstellar medium it inherited the ashes of the nuclear fires that fueled those previous generations of stars. By the Age of the Chemical Elements I mean the time back to when the stars in the Galaxy first began to produce the heavy elements and particularly the radioactive ones. Adding a billion years or so for the time of formation of the Galaxy and for the evolution of its early stars yields a completely independent determination of the Age of the Universe. There is the implicit assumption that all other galaxies formed along with our Galaxy at roughly the same time.

As I stated earlier the final result falls close to 11 billion years. That is an incredible time compared to the lifetime of an individual and even to the time back to the appearance of human beings on earth a few million years ago. For me, it shows the marvelous scope possible in human concepts, and I hope you will share this feeling with me. However, some of you may be skeptical about all this and for you I will quote the great American author and humorist, Mark Twain. In "Life on the Mississippi" Twain wrote "There is something fascinating about science. One gets such wholesale returns of conjecture for such a little investment in fact." Oh, that does strike home!

My work in 1986 on cosmochronology with Chris Meisl, an undergraduate at Caltech, was stimulated by the invention of the Inflationary Model of Cosmology in 1981 by Alan Guth of the Massachusetts Institute of Technology. The Inflationary Model is one of the most elegant things to happen in cosmology in my lifetime. One possible scenario is that our observable universe is just an expanding bubble in a much vaster steady state Universe. After years of controversy between advocates of the old big bang model and the advocates of the old steady state model, Herman Bondi, Tom Gold and Fred Hoyle, it may turn out that they were both right in a way. With the assumption of an enormous inflation in the size of the Universe when it was only 10^{-34} to 10^{-32} seconds old the Inflationary Model is able to explain why astronomers find that the universal background radiation has the same temperature, 2.736 degrees Kelvin above absolute zero, everywhere they look in the sky. It solves the so-called horizon problem which plagues the traditional big bang model in which the early universe was larger than the horizon for a hypothetical observer in any region of the universe. Thus the universe could be divided into regions which were not in casual contact; so how did all these regions come to have the same cosmic background temperature? In the Inflationary Model the early universe, before the enormous inflationary increase in its size, was very small and all regions were in casual contact at the speed of light. The horizon of any hypothetical observer exceeded the radius of the universe. Thus from the basic physics of heat transfer the temperature of the early universe was the same everywhere and has cooled through expansion to the same temperature everywhere today as astronomers find to very high accuracy. At later stages the inflation fits into the big bang model in time to explain the primordial synthesis of the two isotopes of hydrogen, ^{1}H and ^{2}H, the two isotopes of helium, ^{3}He and ^{4}He, and a small fraction of the still heavier elements. Moreover it involves elementary particle physics and high energy physics in our understanding of the early Universe and that involvement promises to be very exciting in the years to come.

For our purposes in this lecture the Inflationary Model predicts that the Universe is Euclidean or "flat" in common parlance. Einstein's curvature of space is zero. The average density of all the matter in the Universe is exactly equal to the so-called critical density which corresponds to about 4 hydrogen atoms in each cubic meter of space or about ten to the minus 23 grams per cubic meter. It then follows that the age of the Universe is exactly two-thirds of the Hubble time. This does require that Einstein's cosmological constant be zero but that is what some modern theories predict. If the cosmological constant is not zero then the vacuum has a density. This strange idea cannot be ruled out completely. In my view the age is indeed close to two-thirds of the Hubble time and in any case the uncertainties in measuring both the age and the Hubble time are such that there is no need to involve a nonzero value for the density of the vacuum. The prediction of the Inflationary Model that the Age of the Universe is exactly two-thirds of the Hubble time is a challenge to all of us who play what I call "The Age Game."

In conclusion it is appropriate to summarize. The Hubble time is close to 17 billion years. The Age of the Universe is 11 billion years give or take two billion years. Thus the Universe expanded more rapidly in the past to reduce the Hubble time of 17 billion years to an age of 11 billion years. The expansion rate has been slowly decreasing.

As I said previously the most elegant theory of the Expanding Universe is the Inflationary Model proposed by Alan Guth and others. According to this theory the Universe contains on average just enough matter-energy to make it what cosmologists call a flat, Euclidean Universe - the simplest of all mathematical possibilities. The cosmological constant is most probably equal to zero. For such a Universe the age is exactly two-thirds of the Hubble time. Eleven over seventeen is close enough to 2/3 for me. I think the Inflationary Model is correct.

I must remind you again. There are those who agree with me and those who do not. The pessimists say it is all so implausible it must be wrong. The optimists say that makes it all the more remarkable if it turns out to be true. There may be some of you who think trying to answer the question "How Old is the Observable Universe?" is a waste of time in view of the large uncertainties. And perhaps you are right. In any case, it is well to remember the wise words which Samuel Pepys entered into his diary on May 23rd, 1661 over 300 years ago. This was during the great controversy in regard to Bishop Ussher's biblical time scale - the Bishop added up all the generations in the Bible and said it all started in 4004 B.C. Against this was the growing realization by geologists that geologic processes on the earth required periods of at least 100 million years. So I quote Pepys in 1661: "To the Rhenish wine-house, and there came Jonas Moore, the mathematician, to us ... and spoke of very many things not so much to prove the Scripture false, as that the time therein is not well computed nor understood." Human beings are fallible in understanding and in computing. There is a lesson there for all of us and especially for me. Anyhow Pepys was a wise old bird so I'll let him have the last word. So be it!

An INFLATIONARY Primer

Joshua W. Burton

Department of Nuclear Physics
Weizmann Institute of Science
Rehovot, Israel 76100

1. Why Inflation?

> *Albert, stop telling God what to do!*
> — Niels Bohr

The aim of this note is the same as that of my talk at Erice: to convey the underlying ideas and motivations behind the inflationary picture, and at least a qualitative sense of its strengths and weaknesses, without invoking either specific particle-theory models or the full general relativistic formalism. In the bibliography I cite several recent books that treat the subject in more detail, as well as a few landmark papers in the original literature; with my handwaving heuristics in the back of her mind, the ambitious reader should be able to approach these without much trepidation.

One basic fact that is often misunderstood about inflation in all its variants is that it is a theorist's solution to *theorists'* problems. In this is differs from, for example, Einstein's postulate of general coordinate invariance or Pauli's hypothesis of the neutrino, both theoretical constructs that explained away well-measured and vexing experimental discrepancies (in the former case, Mercury's perihelion shift; in the latter, the conservation of energy and momentum in nuclear β-decay). Inflation as a solution for the puzzles I will detail below is more like Planck's quantum postulate as a solution to the ultraviolet catastrophe, or like local gauge invariance as an explanation for the masslessness of the photon: the problems it solves are not forced upon us by the data, but rather by apparent failures of our theoretical understanding when we extrapolate it to unexplored regimes. This by no means implies that inflation is useless—as we shall see, it rescues us from at least three theoretical quandaries, and does so in a surprisingly elegant and aesthetic way. Nonetheless, it is well to keep in mind that (as of summer 1992) there is no experimental evidence which directly supports the inflationary view of the early universe. And of course, the universe is not required to answer to our present-day view of what is aesthetically pleasing.

The most venerable of the theoretical problems to which inflation may provide a clue is the so-called *horizon problem*, which is really just a stepchild of Olbers' paradox. Heinrich Olbers realized almost two hundred years ago that the sky of an infinite unchanging universe would not be dark, but rather would have a thermal spectrum at the temperature of the average star. Since this is certainly not what we observe, it follows that the universe is finite in either spatial or temporal extent—in particular, we now believe that the observable universe is constrained by the speed of light and the finite time that has elapsed since the Big Bang. But now a new problem arises: the standard Big Bang cosmology predicts an ever-slowing universal expansion, with the universe growing like $t^{1/2}$ for its first few hundred thousand years, and like $t^{2/3}$ thereafter, where t is the current age of the universe. (The reason for the glitch is that before about $Z = 1000$ the mass of the universe is dominated by relativistic particles

139

M. M. Shapiro et al. (eds.), Particle Astrophysics and Cosmology, 139–146.

which lose energy as they redshift, whereas at later times it is dominated by the rest mass of nonrelativistic matter. The important point for our purposes is that both rates are slower than t itself.) Since the horizon of the observable universe is growing linearly with time, it follows that new regions—apparently never before in causal contact with us—are constantly becoming visible. The question that immediately arises is why such regions so closely resemble the universe we can already observe.

To give this question a more quantitative flavor, consider the photons of the cosmic microwave background. These are thermal photons emitted from excited nuclei in the primordial hydrogen/helium/electron plasma, at a time when all cosmological distance scales were three orders of magnitude smaller than they are today. At about that time, $Z \approx 1500$ or $t \sim 3 \times 10^5$ years, the plasma cooled to the point where it could recombine into neutral atoms, and thereby became nearly transparent. The photons of the CMB have presumably never scattered between that time and the moment when they hit our detectors; their shift from visible wavelengths into the microwave is a result of the overall expansion of the universe, and not of any later interaction with matter. At the time of last scatter for the CMB, the observable horizon was some fifty times smaller than at present, implying that photons reaching us from regions of the sky more than a degree or so apart have never had any chance to exchange thermal information. How then did the CMB become isotropic at the level of a few parts per million?

It is possible to imagine that some initial condition was set at the Big Bang itself, constraining these causally disconnected regions to reach the same temperature at the same time, but this approach leads to further epistemological quagmires. Since as far as we are aware the horizon length has *always* been growing more rapidly than the scale length of the universe, the mutually acausal regions become smaller (even in comoving coordinates) the further back in time we go. If the universe never underwent an inflationary phase, this remains true all the way back to $t = 0$, and the Big Bang must be viewed as the union of infinitely many pointlike noncommunicating regions. In this unpalatable scenario, it is hard to see how the temperature, or indeed any dynamical parameter (including particle masses and mixings, if they arise through symmetry-breaking) can be tuned to the same value everywhere.

A second difficulty with the standard Big Bang cosmology is the flatness problem, or, as Dirac and others termed it, the problem of large numbers. We observe today a universe that is expanding and slowly decelerating, with a characteristic timescale given by the Hubble time, or about 10^{10} years. If the expansion is parabolic, as it would be in a truly flat universe (that is, one whose gravitational binding energy density just cancels its mass density), then this will remain true at all times: the deceleration time will always be precisely two thirds of the current age, and the mass density will always be just enough to balance the remaining binding energy density. We usually refer to such a state of affairs as an $\Omega = 1$ universe, where Ω is defined as the negative ratio of the mass to the binding energy, or equivalently as the ratio of the mass density to the critical density that would be required to recollapse the universe. From direct observation, we know that Ω is bigger than about 0.007, the density of observable luminous matter, and smaller than about 2.0, from the so-called "deceleration parameter" derived from galactic redshift counts. We have good indirect evidence, both from galactic rotation curves and from the proper motions of galaxies in tight superclusters, that the lower bound on Ω must be raised somewhat, certainly to above 0.05 and perhaps to 0.15 or more.

But as we mentioned, Ω evolves with time, with an unstable fixed point at the critical value, one. If Ω is tuned to $1 - \epsilon(t_0)$ at some particular time t_0, it will drift away from one exponentially thereafter, as the universe either recollapses or approaches an asymptotic expansion rate. Reasoning backwards, if we observe $\epsilon(t)$ to be of order unity or less today, it follows that it was smaller than 10^{-4} at the time of photon decoupling, 10^{-16} at the start of nucleosynthesis, and less than 10^{-60} at the Planck time, $t_{Pl} \equiv \sqrt{G_N \hbar / c^5} \sim 5 \times 10^{-44}$ seconds, when such geometric parameters as the universe's total energy density were presumably set. Certainly nothing but taste prevents us from invoking such an extraordinary coincidence; in some sense $1 + 10^{-60}$ is just as "random" a real number as any other. And if the Anthropic Principle, or some similar metaphysical whimsy, compels us to inhabit a universe where our own evolution was possible, then this sort of fine-tuning can even be said to be explicable. Still, it is the business of physics to explain as much as possible in terms of dynamics before appealing to initial conditions, and it would be a relief to see an extremely flat universe emerge in a natural way from the equations of motion*.

Grand unified theories also give rise in general to what is called the monopole problem. Unlike proton decay, this is not a model-dependent artifact of some particular grand unified theory, but rather a general consequence of *any* model that purports to explain the equality of the proton and positron charges by putting quarks and leptons into a single gauge multiplet. Generically, if the $U(1)$ gauge group of electromagnetism is embedded in a semisimple gauge group (as it must be, to explain quantization of charge and the fact that hydrogen is neutral to less than one part in 10^{21}), then the breaking of that larger gauge symmetry should have created approximately one topological defect per horizon volume. These defects will appear as pointlike massive monopoles, typically weighing M_{GUT}/α each, where α is the electromagnetic fine-structure constant. Taking 10^{14} GeV as a typical grand unification energy, we find that the horizon volumes at GUT-breaking have expanded to a meter or so in diameter today; since each monopole weighs 10^{16} GeV$/c^2$ or more (comparable to the mass of a large bacterium), this is a phenomenological disaster. If we don't want primordial monopoles to drastically overclose the universe, we have to either annihilate them or else somehow send them away. Annihilation doesn't seem to work, since the monopole interaction cross-sections are constrained by their known magnetic charges; thus, we are left with inflation as the only viable alternative. This problem was in fact the motivation for the earliest inflation papers, and is still in many ways the most convincing argument for a rapid early expansion, though it is admittedly rather technical and has therefore been somewhat eclipsed by the more intuitive flatness and horizon problems.

Finally, there are more speculative concerns, which can be lumped together as the uniqueness problem. (I am indebted for this term to Collins, Martin, and Squires, cited in the bibliography.) Briefly, why does the universe appear to be four-dimensional at all? If inflation can, as we shall see, explain why the universe appears locally flat, can it

---

* My advisor, Mary K. Gaillard, used to refer to this as the "tooth fairy rule." As she put it, you only get to invoke the tooth fairy once in any given paper. Therefore, it is prudent to explain as much as possible by plausible means, and appeal to fortuitous cancellations and coincidences only where inspiration fails.

possibly explain more general aspects of spacetime geometry and topology? This idea has attracted more attention since the advent of superstring theory, which suggests that the local geometric structure of spacetime may in fact have a higher number of dimensions, such as ten, with the extra dimensions curled up and compactified as in Kaluza-Klein models. If in fact the same mechanism can force some directions to become tightly curved at the same time it makes others relax and become very flat, one process might eventually provide some insight into the other. Some work has been done along these lines, but even what little I understand of it is well outside the scope of the present overview.

2. Inflationary Physics

> *Space is what keeps everything from*
> *happening in the same place.*
> — Dave Barry

Inflation in its modern sense is barely a decade old, but the idea of a universe expanding exponentially with time (rather than in a power law, as in the standard Big Bang) goes all the way back to De Sitter and the dawn of general relativity. Briefly, Einstein's equations of motion allow a source term in addition to ordinary matter and energy, known as the cosmological constant. This constant, if nonzero, causes empty space to behave as though it has a net positive energy density ρ_{vac}, and an equal *negative* pressure p_{vac}/c^2. Since both mass-energy and pressure are self-gravitating, ordinary matter can only slow the universe's expansion rate, but a cosmological constant's peculiar negative pressure will actually accelerate it. The equation governing the expansion rate† is:

$$\frac{\ddot{R}}{R} = -\frac{4\pi G_N}{3}(\rho + \frac{3p}{c^2}) \quad , \tag{1}$$

where $R(t)$ is any comoving distance scale, and ρ and p are the total mass-energy density and pressure, respectively. Clearly, if $\rho = -p/c^2$, the negative pressure will outweigh the positive density, and R will grow exponentially with time. This steady-state solution to the Einstein equations was extensively studied as early as the nineteen-twenties, but it was not until Guth's 1981 paper that it was applied to a realistic cosmological model.

A momentary detour may clear up some confusion about the counterintuitive notion of negative pressure. Naively, it seems that equation (1) is just backwards: in our common experience, it is positive pressures that make things expand, and negative (relative) pressures that hold their expansion in check. But the pressure contribution to equation (1) represents not the pressure itself, but rather the gravitational effect of the pressure—a purely relativistic effect that, because of the c^2, is significant only for pressures so high that the speed of sound becomes comparable to the speed of light. Only for matter at nuclear densities, as for example inside a pulsar, is the pressure correction to Newtonian gravity noticeable. If, however, it were possible to

---

† I had hoped to get through this discussion without any equations at all, in keeping with Hawking's well-known dictum. However, if every equation in a book does in fact halve the potential readership, then all the readers of this volume will have been scared off long before reaching my contribution. Is anyone still here?

create a localized region of negative pressure, that region would rapidly collapse as seen from outside (the reverse of a gas under positive pressure expanding) and would simultaneously cause the space it occupies to expand exponentially. Thus a sort of wart or bubble would form, growing in radius as it is pinched off in circumference. The subsequent evolution of the wart is effectively decoupled from the rest of the universe, and in fact resembles that of an independent inflationary universe; the fascinating possibility therefore arises of creating baby universes in the laboratory! Guth and others have lately considered such questions as what information can in principle pass through the "neck" of the wart from the old universe to the new and vice versa, but such speculations verge rather uncomfortably on the theological.

The simplest example of a concrete particle-physics model that actually causes inflation is a scalar field with a nonzero vacuum expectation value at zero temperature. At very high temperatures, the field sees an effective potential that is symmetric about the origin, but as the temperature falls a true minimum develops at the zero-temperature vev. If the cooling is rapid enough, the field can become trapped in the symmetric state (just as supercooled water can become trapped in the liquid state), and quantum tunneling to the true vacuum can then create nucleation sites for the phase transition to the nonsymmetric state. During the period of supercooling, the vacuum energy of the scalar becomes much larger than its average kinetic energy, and this vacuum energy appears in the gravitational equations of motion as a local cosmological constant. If this constant is large compared to all other gravitational sources, the symmetric regions of the universe will begin to expand with positive $\dot{R}/R$; that is, exponentially.

As I mentioned earlier, inflation was first considered as a solution to the monopole problem, in the context of grand unified models. Thus the phase transition considered was from a GUT-symmetric vacuum to one in which the gauge group was broken to $SU(3)_C \otimes SU(2)_L \otimes U(1)_Y$. However, any sufficiently energetic phase transition will do, provided that it takes place no earlier than the GUT-breaking transition. An earlier inflation, of course, would not solve the monopole problem, since an average of one monopole per horizon volume would be produced *after* the expansion had halted.

3. Inflationary Cosmology

You'll grow into it!

— my grandmother

Guth's original inflationary model, outlined above, is probably unacceptable as a cosmology for the observed universe. In particular, the unbroken phase continues to inflate exponentially, while any bubbles of broken phase that may nucleate rapidly roll down to the true vacuum state and resume ordinary $R \propto t^{1/2}$ expansion. Unlike the situation in phase transitions of water, the bubbles of new phase are being blown away from each other by the intervening old phase more rapidly than they are expanding, and so the universe never becomes homogeneous. Even if the bubbles could eventually coalesce, they formed out of causal contact with each other, and therefore will in general have different phases of true vacuum. The resulting defects, such as domain walls and cosmic strings, would retain the huge energy density of the unbroken state, and their gravitational influence would prevent a locally smooth observable universe like ours from evolving.

Linde, Albrecht and others (in keeping with the informal character of this note, I am leaving the references to the end) found a way out of these problems about a year later. If the scalar potential is flat enough around its maximum, the field will roll over only slowly, and its energy will continue to be dominated by the vacuum potential for some time after symmetry-breaking. In this way, the bubbles of broken phase can continue to inflate for many e-foldings before the scalar settles in the true minimum and reheats the universe by dissipative interactions with matter. This scenario can in principle solve all of the cosmological problems we discussed in the first section.

Monopoles were formed either before or during the inflationary phase transition, but during inflation each horizon volume expanded enormously, so that the entire observable universe today could easily have come from a small fraction of a single GUT-era horizon volume. Rather than being swamped with primordial monopoles, we then expect to have a hard time finding even one within our observable horizon. (Sixty-five years ago, Dirac pointed out that a single monopole anywhere in the universe could explain the quantization of electric charge. It would be very amusing if nature has parsimoniously given us only the one magnetic monopole we really need.) The flatness of the observed universe is explained even more directly by inflation: any curvature that our inflating bubble may have has been diluted by the expansion. If inflation went on for as many as sixty e-foldings, then the bubble is flat enough to have washed out all $\Omega \neq 1$ information up to the present day. If inflation went on for any longer than that, then Ω will still be indistinguishably different from one today. Strictly, then, inflation's famous prediction of critical density has a loophole, but only if we fine-tune to allow *just* enough inflation to occur. Since $\rho = \rho_{crit}$ is certainly not ruled out by observation (though much of that density ρ must be nonbaryonic to avoid excessive helium production in primordial nucleosynthesis), this loophole is generally dismissed as unaesthetic.

Finally, the horizon problem can also be inflated away. Everything we see today was much closer together before inflation than the standard Big Bang model would suggest. Opposite sides of our sky (and indeed, regions that are now causally inaccessible to us) had time to thermalize in causal contact, then were blown out of sight of each other in the inflationary era, and are only now coming slowly back into view, with identical temperatures as souvenirs of their earlier mutual encounter. This notion of regions falling out of causal contact seems profoundly counterintuitive, since it means that they are receding from each other faster than light can bridge the gap. The point is that special relativity only constrains relative speeds to be less than c in any one local inertial frame; in a curved universe two locally flat regions can participate in superluminal relative expansion without actually moving at all. Rather, it is the space between the two reference points that is growing, while each point is constrained to move only subluminally in its own locally Minkowskiian coordinate system. In some sense, information *is* being conveyed faster than c, but it has no choice about whether or not to go! And since the expansion is global and irreversible, there is no way to create closed timelike loops and other causal paradoxes.

Since inflation (in its simplest form) continues for at least long enough to make all preexisting distance scales unobservably large, the fluctuations that will later actually nucleate structure formation in the observable universe must be quantum fluctuations left over from inflation itself. One very general prediction of inflation is that these fluctuations will be scale-invariant; a spectrum of fluctuations that exhibited a preferred distance scale would be very hard to explain if all scales have been inflated away. Recent observations by COBE and balloon experiments of slight fluctuations in the microwave background appear consistent with this prediction; it will be interesting to see whether this remains true when better data are available.

Has inflation solved anything? Certainly, it has eliminated some very unnatural fine-tuning of the universe's initial conditions, but if it requires fine-tuning of the scalar equations of motion to implement, it may be that nothing has been gained. In particular, the extremely flat "hilltop" in the scalar potential (required for inflation to continue for many e-foldings after the slow rollover begins) looks suspiciously fine-tuned: while it is true that flat Coleman-Weinberg scalar potentials arise naturally in finite temperature field theories, a potential flat enough to allow sufficient inflation requires among other things a coupling constant of order 10^{-12}. Fortunately, there is more than one way to stop a rock from rolling down a hill. The top of the hill may be flat, as in the minimal inflation model we have outlined, or alternatively, there may be friction slowing the rock's slide to the bottom (this can be implemented by coupling the scalar to a fermion condensate, so it dissipates its vacuum energy making fermion-antifermion pairs). The rock may simply have started very high on the hill; this is the situation in chaotic inflation models, where a random amount of inflation takes place at each point, and we happen to be in a much-inflated region merely because such regions are now vastly bigger than the others. Or, finally, the rock may be a spinning top and not a rock: I have done some work recently on a model of "spiralling inflation" where the scalar that inflates the universe has a self-interaction which causes it to lose its way whenever it starts downhill, much as a gyroscope converts its potential energy into precessional motion.

The inflationary model is both simple enough to be theoretically appealing and general enough that a decade of study has not exhausted its viable implementations. Advances in particle physics and observational astrophysics in the coming decades may tell us whether it also has any real predictive power.

4. Bibliography

He who quotes his sources brings deliverance to the world.
— Rabbi Meir, quoting Rabbi Yehoshua ben Levi

I felt that a list of pointers for further reading would be more useful than a formal reference list; here are a few of the sources that have shaped my understanding of the inflationary picture.

Keith A. Olive, *Phys. Rep.* **190**, 307 (1990). An excellent and fairly up-to-date review article, with an ample historical overview. Over 500 papers in its reference list!

P. Collins, A. Martin, and E. Squires, *Particle Physics and Cosmology*, John Wiley and Sons (1989), ISBN no. 0-471-60088-1. A very well-written text, with a clear one-chapter treatment of inflationary physics and cosmology. Perhaps the best one-volume reference on modern particle astrophysics now available.

I'll also cite a few of the most important primary references. These will give you a sense of how the subject has evolved over time.

A. H. Guth, *Phys. Rev.* **D23**, 347 (1981). The paper that started the industry.

A. D. Linde, *Phys. Lett.* **B108**, 389 (1982), and A. Albrecht and P. J. Steinhardt, *Phys. Rev. Lett.* **48**, 1220 (1982). The beginning of new inflation, and the slow-rollover scenario.

A. D. Linde, *Phys. Lett.* **B129**, 177 (1983). Chaotic inflation, and non-thermal initial states.

L. F. Abbott and M. B. Wise, *Nucl. Phys.* **B244**, 541 (1984). A early look at the possibility of non-exponential inflation, now an important notion in so-called extended inflation models.

A. H. Guth and S.-Y. Pi, *Phys. Rev.* **D32**, 1899 (1985). The quantum mechanics of the inflationary slow rollover.

G. F. Mazenko, W. G. Unruh, and R. M. Wald, *Phys. Rev.* **D31**, 273 (1985). A more general discussion of quantum mechanics and inflation, cataloguing classical ambiguities and global aspects requiring a quantum treatment.

I would like to acknowledge the support of the Sir Charles Clore Postdoctoral Fellowship and the Mozart G. and Eliot Ratner Memorial Fellowship, which enabled me to participate in this school, and of course to join my fellow lecturers and students in thanking the organizers for their time and effort.

FLUCTUATIONS IN THE COSMIC BACKGROUND RADIATION

M. Giller, A.W. Wolfendale,
Physics Department,
University of Durham,
South Road,
Durham, DH1 3LE, UK.

ABSTRACT. The search for fluctuations in the CMB is one of the major challenges of contemporary Astrophysics. Not surprisingly, the sought-after signal is affected by foreground effects generated in our own Galaxy. The present work concerns the likely magnitude of these effects and their influence on the conclusions to be drawn from the recent observations of the CMB, most notably those from the COBE satellite.

1. Introduction

Although the observation of the CMB radiation was made over 25 years ago it is only recently that the observations have become sufficiently precise for claims to have measured the fluctuations in this radiation to be taken seriously. The problems are well known - the effect sought is very small, $\Delta T/T \approx 10^{-5}$ on a 10° scale, and the undoubted presence of fluctuations of Galactic origin and due to Galactic synchrotron radiation (GSR) and Galactic dust (GD). The foreground effects are clearly frequency dependent, GSR dominating at low frequencies and GD at high frequencies ('high' and 'low', here, refer to a datum of $\simeq 100$ GHz). In what follows we summarise the situation with respect to GSR and GD.

2. Ground Level Studies at Low Frequencies

At radio frequencies below 10 GHz the sky is dominated by GSR, viz synchrotron radiation generated by cosmic ray electrons gyrating in the Galactic magnetic field. There have been measurements of precision at frequencies of 408, 820 and 1420 MHz and these have been used by a number of authors (e.g. Banday and Wolfendale, 1991 and Banday et al., 1991) to derive the fluctuations expected at higher frequencies. The procedure, by extrapolation of the lower frequency data, is not trivial because of uncertainties in the spectral index of GSR and its dependence on l,b. Nevertheless, some efforts have been made and Figure 1 shows the predicted '$\Delta T/T$' (actually the rms second difference divided by 2.73 K) as a function of frequency. The angular

M. M. Shapiro et al. (eds.), Particle Astrophysics and Cosmology, 147–151.

scale in question is 5.6°. It will be noted that measurements below 20 GHz will certainly be strongly affected by GSR if, as seems likely for other reasons, the genuine CMB fluctuations are a little below 10^{-5}. It should be stressed that the predictions relate to a region of the sky: $\delta \sim 40°$, RA : 180° - 250°, for which the GSR signal is almost certainly a minimum.

Also shown in Figure 1 is the observed value of σ_s/T from the work of the Jodrell-Cambridge-IAC group at Tenerife (Rebolo et al., 1989). These workers had put forward their signal as a candidate CMB fluctuation but it is apparent that it can equally well be explained by GSR. There is a problem, however, in that the actual profile of ΔT versus RA is not as predicted by extrapolation from the lower frequencies; it is necessary to postulate changes of spectral index with position as the cause.

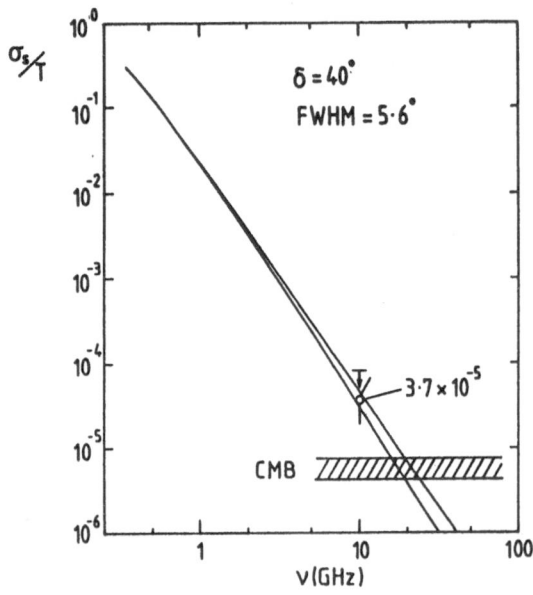

Figure 1 Predicted GSR anisotropy vs. frequency. The values relate to the rms of second-difference ΔT_s divided by $T = 2.7$ K, denoted as σ_s/T, and arise from an extrapolation of the measured T_{408} and T_{1420} maps. The results refer to $\delta \simeq 40°$, R.A.: 180° - 250°. Also shown is the signal at 10.46 GHz given by Davies et al. (1987) and the upper limit from Rebolo et al. (1989). The CMB prediction for the isocurvature model is also shown.

Figure 2 shows the situation for the most recent radio observations of Watson et al., (1992) at 14.9 GHz. Here, we present an upper limit to σ_s/T as quoted by Watson et al; it is seen to be quite consistent with our prediction. There is, however, again a problem with GSR - the observed pattern of ΔT versus RA at the higher frequency does not correlate with that at 10.46 GHz. It is not at all clear whether this means that instrumental noise dominates at the higher frequency or that the radio spectrum is very steep - an exponent significantly greater than the commonly adopted 3.0. This problem really does need solving before radio measurements at frequencies below about 30 GHz are used to determine the bona fide CMB fluctuations.

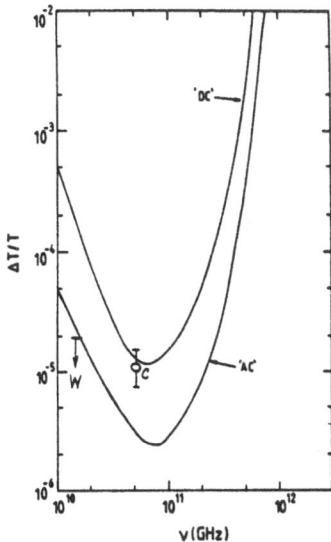

Figure 2 Our best estimate of the Galactic contribution to $\Delta T/T$ for a number of frequencies. The line marked DC corresponds to the situation at b = 90° derived by extrapolating from low-frequency radio surveys and from the IRAS 100-μm survey (assuming the Lubin et al. spectrum). The AC level corresponds to first-difference measurements of $\Delta T/T$ for COBE in a hypothetical 'quietest region' (deduced again from the radio and 100-μm maps). The upper limit from the Watson et al. (1992) observations is indicated W, and the actual COBE result from Smoot et al. (1992) averaged over all Galactic latitudes above 20°, is shown as C.

Returning to the Watson et al. data, we have derived σ_s from the 14.9 GHz results in two ways. Firstly, the contribution from GSR is ignored; the values then are 46 ± 8 (5), 20 ± 8 (10) and 14 ± 8 (20), the values being in μK and the number in brackets being the number of degrees of RA over which the averages are taken. For example the COBE results should be compared with (10); for this size of 'cell' we expect a value of 30 μK from noise alone and the genuine CMB signal would be zero at the one standard deviation level, and 26 μK at the 2σ level. The corresponding value of σ_s/T is 1 x 10^{-5} at the 2σ level.

If, now, it is assumed that GSR is present at 14.9 GHz and that its magnitude is as given by the 10.46 GHz signal attenuated with a spectrum having exponent 3.0, the overall genuine 'signal' for 10° increases from 20 μK to 57 ± 13 μK. The expected noise is also higher: 47 μK and the signal less noise is higher. Specifically, the best estimate is $\sqrt{57^2 - 47^2} = 32$ μK (± 34 μK), corresponding to $\sigma_s/T = (1.2 ± 1.3)10^{-5}$.

3. The COBE Observations

It is clear from Figures 1 and 2 that GSR becomes less important at higher frequencies and inevitably observations from space are called for. Such measurements have been made recently by the COBE group (Smoot et al., 1992). These authors have observed on a 7° scale at frequencies of 31, 53 and 90 GHz. Two radiometers, A and B, observing at angles separated by 60°, were used to provide $\frac{A+B}{2}$, comprising signal + noise and $\frac{A-B}{2}$ comprising noise alone.

Table 1 summarises the results. It will be noted that the 'signal', $\sigma(_{sky})$, is derived by subtraction of $\sigma(\frac{A-B}{2})$ from $\sigma(\frac{A+B}{2})$ in quadrature. The authors argue that the near constancy of σ_{sky} for Galactic latitudes above 20° confirms their contention that the foreground effect are negligible and that $\sigma_{sky}/T \simeq 10^{-5}$.

ν	b_o	σ_{obs} $\frac{A+B}{2}$	Noise $\sigma \frac{A+B}{2}$	σ_{sky}
31 GHz	10°	133	94	$95\pm^{6}_{6}$
	20°	101	96	$33\pm^{9}_{12}$
	30°	97	95	$20\pm^{21}_{20}$
	40°	101	96	$30\pm^{12}_{15}$
53 GHz	10°	55	36	$41\pm^{4}_{4}$
	20°	46	35	$30\pm^{5}_{5}$
	30°	47	36	$30\pm^{5}_{5}$
	40°	47	36	$30\pm^{5}_{5}$
90 GHz	10°	69	59	$35\pm^{6}_{7}$
	20°	65	61	$22\pm^{7}_{10}$
	30°	67	60	$29\pm^{7}_{8}$
	40°	67	59	$31\pm^{7}_{8}$

<u>Table 1</u> RMS values for 10° smoothing for various latitude cuts (b $\succ$ b$_o$) from the COBE experiment of Smoot et al. (1992).

Our own estimate of the foreground to be expected at these frequencies can be seen in Figure 2, taken from Banday and Wolfendale (1991). It will be noted that our estimate of the foreground fluctuations in the second differences (denoted 'AC') is only a little above 2×10^{-6} at the two upper COBE frequencies and at first sight it would appear that 'all is well'. However, the predictions are for high Galactic latitudes and we would expect the result for b $\succ$ 20° to be significantly higher. Furthermore, no allowance was made by us for free-free emission which would in fact be expected to be significant, compared with GSR, at frequencies above about 30 GHz.

The basic approach to searches for genuine fluctuations in the CMB is to identify a component with the correct spectral shape, viz, $\Delta T/T$ should be independent of frequency. If there were only GSR or free-free or GD there would be no problem; however, as can be seen from Figure 2 in the region of the minimum of 'AC' the sum of the components gives a nearly frequency - independent $\Delta T/T$ and the test fails.

4. Conclusions

Although COBE appears to have detected bona fide CMB fluctuations care is needed because of the possibility of there being a still non-negligible contribution from foreground effects. Although the results are probably correct (and there is perhaps a little support - certainly no contradiction, yet, - from the 14.9 GHz results), it would be wise not to rely too heavily on detailed aspects of the data until a more thorough study has been undertaken and independent observations made. Of particular need is maps of the (claimed) CMB sky so that the hot spots, which contribute so much to the derived rms value of ΔT, can be identified. Searches for alternative sources of the higher temperatures can then be made. The situation at present is that, as Smoot et al. (1992) and Bennett et al. (1992) point out, the published maps have very considerable contributions from 'noise'. 'Genuine' maps are needed.

REFERENCES

Banday, A.J. and Wolfendale, A.W., (1991) 'Galactic Dust Emission and the Cosmic Microwave Background', Mon. Not. R. Astr. Soc. 252, 462-472.

Banday, A.J., Giller, M., Szabelska, B., Szabelski, J. and Wolfendale, A.W., (1992), 'Cosmic Rays and Cosmological Microwave Background Fluctuations', Astrophys. J., 375, 432-438.

Bennett, C.L. et al., (1992) 'Preliminary Separation of Galactic and Cosmic Microwave Emission for the COBE-DMR', Astrophys. J. (in press).

Smoot, G.F. et al., (1992) 'Structure in the COBE DMR First Year Maps', Astrophys. J. (in press).

Rebolo, R., Watson, R. and Beckman, J.E., (1989). Ap. Space Sci. 157, 333-340.

Watson, R.A., Gutierrez de La Cruz, C.M., Davies, R.D., Lasenby, A.W., Rebolo, R., Beckman, J.E. and Hancock, S., (1992) 'Anisotropy Measurements of the Cosmic Microwave Background Radiation at Intermediate Angular Scales', Nature, 357, 660-665.

PARTICIPANTS

Tal ALEXANDER
Department of Astronomy
School of Physics and Astronomy
Tel Aviv University
Tel-Aviv 69978
ISRAEL
E-Mail: tal@wise3.tau.ac.il

Pietro ANTONIOLI
Universita degli Studi di Torino
Istituto di Fisica Generale
Via Pietro Giuria, 1
10125 Torino, ITALY
E-Mail: antonioli@vaxto.infn.it
-or- 39278::antonioli

Hussein BADRAN
Department of Physics
Tanta University
Tanta
EGYPT

Evgeny BEREZHKO
Inst. of Cosmophysical Research
Lenin Ave 31
Yakutsk 677891
RUSSIA
Telex: 145126 lena su

Konrad BERNLHOR
Max-Planck-Institut
fur Kernphysik
Postfach 103980
D-6900 Heidelberg 1
GERMANY
E-mail: bernlohr@eu6.mpi-hd. mpg.de

Stella Marie BRADBURY
Department of Physics
University of Durham
South Road
DH1 3LE Durham
UNITED KINGDOM
E-mail: sbradbury@star.dur.ac.uk

Joshua BURTON
Dept. of Nuclear Physics
The Weizmann Institute of Science
Rehovot 76100
ISRAEL
E-mail: ftburton@weizmann.ac.il
-or- burton@QCD.physics.brown.edu

Robert CALDWELL
Department of Physics
University of Wisconsin
P.O. Box 413
Milwaukee, WI 53201 USA
E-mail: caldwell@dirac.phys.uwm. edu

Sylvie CHARBIT
C.E.N. Saclay
DAPNIA/SPP bat. 141
91191 Gif sur Yvette Cedex
FRANCE
E-mail: charbit@frsac53.bitnet -or-
dphvx2::charbit

Arati CHOKSHI
IPAC: 100-22
California Institute of Technology
Pasadena, CA 91125
USA
E-mail: Chokshi@IPAC.Caltech.edu

John CLEM
Department of Physics & Astronomy
Louisiana State University
Baton Rouge, LA 70803-4001 USA
E-mail: 7894::clem -or-
clem@phepds.phys.lsu.edu

154

James CONNELL
Laboratory for Astrophysics and Space
Research
The University of Chicago
933 E. 56th Street
Chicago, IL 60637 USA
E-mail: lasr::connell -or-
connell@odysseus.uchicago.edu

Kevin DESOUZA
Department of Physics
University of Leeds
Woodhouse Lane
LS2 4JT Leeds
UNITED KINGDOM
E-mail: Phy5kds@sun.leeds.ac.uk

Lev DORMAN
Department of Cosmic Ray
Izmiran
142092 Troitsk
RUSSIA
E-mail: leva@ve.tau.ac.il

Georgia DRILIA
Department of Physics
University of Athens
Panepistimiopolis
157 83 Zografos
Athens, GREECE
E-mail: (span): 5417::Drilia

Debiprosad DUARI
Inter-University Centre for Astronomy
and Astrophysics
Post Bag 4
Pune 411007
INDIA
E-mail: debi@iucaa.ernet.in

Nevin DULGER
Department of Physics
Erciyes University
Talas Yalu
38039 Kayseri
TURKEY
E-mail: dulger@trerun.bitnet

Seyed FATEMI
Department of Physics
University of Kerman
P.O. Box 76175-133
Kerman
IRAN
Fax: 0341-42716

Giovanni FAZIO
Optical Infrared Division
Harvard Smithsonian Center for
Astrophysics
60 Garden Street
Cambridge, MA 02138 USA
E-mail: fazio@cfa.harvard.edu

Ervin FENYVES
Department of Physics
P.O. Box 830688
University of Texas at Dallas
Richardson, TX 75083 USA

David FICENEC
Department of Physics
Washington University
Campus Box 1105
One Brookings Drive
St. Louis, MO 63130-4899 USA
E-mail: ficenec@landau.wustl.edu

William FOWLER
Dept. of Phys. Math. & Astron.
California Institute of Technology
1225 California Boulevard
Pasadena, CA 91125
USA
E-mail: keay@caltech.bitnet

Peter GABRIEL
Institut fur Kernphysik
der Universitaet Karlsruhe
Kernforschungszentrum
Postfach 3640
W- 7500 Karlsruhe 1
GERMANY
E-mail: Ikpoc4@Dkakfk3.bitnet

Emmanuelle GAILLARD
CEA-Saclay
DAPNIA/SPP, bat 141
91191 Gif sur Yvette Cedex
FRANCE
E-mail: Gaillard@frsac53.bitnet
-or- Dphvx2::Gaillard

Franco GIOVANNELLI
Consiglio Nationale delle Richerche
Istituto di Astrofisica Spaziale
C. P. 67, via E. Fermi, 21
00044 Frascati (Rm)   ITALY
E-mail: flavio@irmias.bitnet

Carlo GIUNTI
INFN
Via P. Giuria 1
10125 Torino   ITALY
E-mail: Vaxto::giunti -or-
giunti@torino.bitnet.infn.it

Lilian GRAHAM
Department of Physics
Science Laboratories
South Road
DH1 3LE Durham
UNITED KINGDOM
Fax: 44-91-374-3749

Stephen GRANT
Department of Physics
Imperial College
Prince Consort Road
South Kensington
SW7 2BZ London
UNITED KINGDOM
E-mail: smg@star.ic.ac.uk -or-
19759::SMG

Stephen HUBER
Beaver College
Dept. of Chemistry & Physics
Glenside, PA 19038-3295 USA
Fax: 215-572-0240

Einar JULIUSSON
Department of Physics
Science Institute
Dunhaga 5
Reykiavik
ICELAND

Frank KRENNRICH
Max-Planck-Institute fur Physik
Fohringer Ring 6
D 8000 Munchen 40
GERMANY
E-mail: 14037::frank -or-
Frank@hegra1.mppmu.mpg.de

Denis LEAHY
Dept. of Physics & Astronomy
The University of Calgary
2500 University Drive NW
Calgary Alberta T2N 1N4
CANADA
E-mail: Leahy@IRAS.ucalgary.ca

Jeremy LLOYD-EVANS
Department of Physics
University of Leeds
LS2 9JT Leeds
UNITED KINGDOM
Fax: 0532 333 846

Rita LOMBARDI
Istituto di Astronomia
Facolta di Scienze MFN
Universita di Roma "La Sapienza"
Via G.M. Lancisi 29
00161 Roma,   ITALY
Fax: 06-4403673

Nathaniel LONGLEY
Dept. of Physics & Astronomy
Carleton College
One North College Street
Northfield, MN 55104 USA
E-mail: mnhep1::nat -or-
nat@mnhep1.hep.umn.edu

156

Marvin MARSHAK
School of Physics and Astronomy
University of Minnesota
116 Church Street SE
Minneapolis, MN 55455 USA
E-mail: mnhep1::marshak -or-
marshak@mnhep1.hep.umn.edu -or-
mlm@anlhep.bitnet

Giulio MIGNOLA
INFN Sezione di Torino
Via P. Giuria 1
10125 Torino, ITALY
E-mail: mignola@torino.infn.it.
bitnet

Richard MILLER
Dept. of Physics and Astronomy
Louisiana State University
Baton Rouge, LA 70803-4001 USA
E-mail: 7894::Miller -or-
miller@phepds.dnet.nasa.gov

Ken'ichi NOMOTO
Department of Astronomy
University of Tokyo
Bunkyo-ku
113 Tokyo, JAPAN
E-mail: nomoto@apsun1.astron.s.
u-tokyo.ac.jp -or- 41963:: nomoto

Mehmet OZEL
Space Science Department
Marmarer Res. Center of Tubitak
41470 Gebze-Kocaeli
TURKEY
Fax: 199-12309

Jose Ramon PELAEZ
Departamento de Fisica Teorica I
Facultad de Ciencias Fisicas
Universidad Complutense
28040 Madrid
SPAIN
E-mail: Jpelaz@emducm11.bitnet

Christoph PFEIFER
Department of Physics
Universitat Siegen
Fachbereich 7
Adolf-Reichwein-Strasse
5900 Siegen, GERMANY
E-mail: Dsiva::Hesse

Martin POHL
Max-Planck-Institut fur Radioastronomie
Auf Dem Hugel 69
5300 Bonn 1
GERMANY
E-mail: P578pom@Mpirbu.mpg.de

Raymond PROTHEROE
Dept. of Physics and Math Physics
The University of Adelaide
GPO Box 498,
5001 Adelaide, AUSTRALIA
E-mail: RProther@physics.adelaide.
edu.au

Francois QUEINNEC
CEN/Saclay
DAPNIA/SPP, Bat. 141
91191 Gif-sur-Yvette Cedex
FRANCE
E-mail: Queinec@frsac53.bitnet
-or- Dphvx2::Queinec

Sisir ROY
Indian Statistical Institute
203 Barrackpore Trunk Road
Calcutta 700 035
INDIA
E-mail: Sisir@Isical.ernet.nest.in

Jalal SAMIMI
Department of Physics
Sharif University of Teheran
P.O. Box 11365-9161
11365 Teheran
IRAN
Fax: 98-21-401-2983

Harald SCHIELER
Institut fur Kernphysik
der Universitaet Karlsruhe
Kernforschungszentrum
Postfach 3640
W-7500 Karlsruhe 1
GERMANY
E-mail: schieler@ikl.kfk.dbp.de

Eun-Suk SEO
Institute for Physical Sciences and
Technology
University of Maryland
College Park, MD 20742-2431
USA
E-mail: ES83@umail.UMD.EDU

Maurice SHAPIRO
205 Yoakum Pkwy. #1514
Alexandria, VA 22304
USA

Yakov SHNIR
Academy of Sciences of Belarus
Laboratory of Theoretical Physics
F. Skaryna Avenue 70
Minsk 220602
Republic of Belarus
E-mail: kilin@adonis.ias.msk.su

Rein SILBERBERG
Code 4154
Naval Research Laboratory
Washington, DC 20375
USA
E-mail: crs1::tsao

Malabika SINHA
Indian Statistical Institute
203 Barrackpore Trunk Road
Calcutta 700035
INDIA
E-mail: Sisir@Isical.ernet.nest.in

Raghunathan SRIVATSAN
HECR Group
Tata Institute of Fundamental Research
Homi Bhabha Marg, Colaba
Bombay 400 005 INDIA
E-mail: Srivat@tifrvax.bitnet

Todor STANEV
Bartol Research Institute
University of Delaware
Newark, DE 19716-4793
USA
E-mail: Bartol::Udbri::Stanev -or-
Stanev@Bartol.Bartol.del.edu

Christopher STARR
Code 4153
GRO Science Support Center
Naval Research Laboratory
4555 Overlook Ave. SW
Washington, DC 20375 USA
E-mail: Starr@Grossc.gsfc.nasa.gov
-or- 11335::Starr

Bertram STILLER
1870 Wyoming Ave., NW
Washington, DC 20009
USA

Alexei STRUMINSKY
Lebedev Physical Institute
Leninsky Prospect 53
Moskow, RUSSIA
E-mail: Stozhkov@Sci.fian.msk.su

Anthony SZABO
Dept. Physics & Math. Physics
University of Adelaide
GPO Box 498
5001 Adlaide
AUSTRALIA
E-mail: Aszabo@physics.adelaide.
edu.au

Yasuo TANAKA
Institute of Space and Astronautical
Science
3-1-1 Yoshinodaz
229 Sagamihara
JAPAN

Klaus THIELHEIM
Institute fur Reine und Angewandte
Kernphysik
Universitat Kiel
Otto Hahn Platz 1
D-2300 Kiel 1
GERMANY

Eric TORBET
Enrico Fermi Institute/LASR
University of Chicago
933 East 56th Street
Chicago, IL 60637 USA
E-mail: Torbet@Odysseus.uchicago.edu

Bulent UYANIKER
Department of Physics
Middle East Technical University
06531 Ankara
TURKEY
Fax: 4-223-6945

Mayank VAHIA
Cosmic Ray/Space Physics
Tata Institute of Fundamental Research
400005 Bombay
INDIA
E-mail: Vahia@Tifrvax.bitnet

John WEFEL
Dept. of Physics & Astronomy
Louisiana State University
202 Nicholson Hall
Baton Rouge, LA 70803-4001
USA
E-mail: 7894::Wefel -or-
 Wefel@Phepds.dnet.nasa.gov

Lance WILLIAMS
Dept. of Planetary Science
University of Arizona
Tuscon, AZ 85721
USA
E-mail: Lance@Hindmost.
LPL.Arizona.edu

Arnold WOLFENDALE
Department of Physics
Durham University
South Road
DH1 3LE Durham
UNITED KINGDOM

Aylin YAR
Physics Department
Middle East Technical University
06531 Ankara
TURKEY
E-mail: A10380@Trmatua.bitnet

Ran ZAMIR
Physics Department
Ben-Gurion University
Beer-Sheva 84105
ISRAEL
E-mail: Ran@Bguvms.il.bitnet

AUTHOR INDEX

164